The New World of Mr Tompkins
物理世界奇遇记

〔美〕G. 伽莫夫 〔英〕R. 斯坦纳德 著

刘 兵 评点

科学出版社

北 京

图字：01-2004-3929 号

The New World of Mr Tompkins, 1st edition (ISBN: 0-521-63009-6) by George Gamow and Russell Stannard, first published by Cambridge University Press, 1999. All rights reserved.

This reprint edition for the People's Republic of China is published by arrangement with the Press Syndicate of the University of Cambridge, Cambridge, United Kingdom.
©Cambridge University Press & Science Press, 2006

This book is in copyright. No reproduction of any part may take place without the written permission of Cambridge University Press or Science Press.

This edition is for sale in the People's Republic of China (excluding Hong Kong SAR, Macao SAR and Taiwan Province) only.

此版本仅限中华人民共和国境内（不包括香港、澳门特别行政区及中国台湾）销售。

图书在版编目(CIP)数据

物理世界奇遇记 /（美）伽莫夫（Gamow, G.），（英）斯坦纳德（Stannard, R）著；刘兵评点. —北京：科学出版社，2006
（20 世纪科普经典特藏）
ISBN 978-7-03-016649-4

Ⅰ. 物⋯ Ⅱ. ①伽⋯②斯⋯③刘⋯ Ⅲ. 物理学–普及读物 Ⅳ. 04–49

中国版本图书馆CIP数据核字（2005）第 152541 号

责任编辑：胡升华 郝建华 / 责任校对：李弈萱
责任印制：李 彤 / 封面设计：黄华斌

科学出版社 出版
北京东黄城根北街 16 号
邮政编码：100717
www.sciencep.com
北京建宏印刷有限公司 印刷
科学出版社发行 各地新华书店经销
*
2006 年 2 月第 一 版　开本：720×1000 1/16
2022 年 6 月第十七次印刷　印张：17 1/4
字数：338 000

定 价：49.00 元
（如有印装质量问题，我社负责调换）

序

20世纪在科学发展史上是一个辉煌的世纪，以物理学和生物学的创新性成果为标志的科学成就，极大地改变了世界的面貌，改变了人类的认知水平、生产方式和生活方式。20世纪也是科学史上的一个英雄世纪，一大批别具一格的科学大师风云际会，相继登场，使科学的舞台展现出前所未有的绚丽风采。20世纪发生了两次世界大战，二战催生的原子弹，使社会公众了解了科学的巨大威力，也促使人类认真地审视科学，了解到科学必须要与人类的良知，与人文精神结合在一起，只有合理地利用，才能造福于人类，才能有利于和平，有利于人类社会的可持续发展。进入20世纪80年代，人类更进一步认识到必须携起手来保护生态，控制环境污染，探索可持续发展的道路。可持续发展理念的形成，是20世纪阶级社会发展观进步的一个重大的事件。

回顾20世纪科学走过的道路，从突飞猛进的科学创造，到科学与人文伦理的深度撞击，形成与人文精神交融并进的局面，最终在人类文明史上留下了不同寻常的篇章。

20世纪诞生的科学和思想大师所取得的非凡的科学成就、创造的充足科学和思想养分，孕育了一批优秀的科普作品，为公众提供了丰富的精神食粮。人们可以跟着爱因斯坦、薛定谔、伽莫夫、沃森、温伯格、霍金等等科学大师的生花妙笔去领略科学创造的历程、登攀一个个科学顶峰的征程和科学高峰的神奇景观；可以跟着卡逊在寂静的春天里思考知更鸟的命运；可以跟着萨根去观察宇宙和生命……。今天这些科学大师和思想大师大部分都已离开了我们，但那些优秀科普作品是他们留给后代的不朽的精神财富。

20世纪已经过去，21世纪已经肯定是一个全球化、知识化的世纪，也是科技国际化、网络化的一个时代。可持续发展依然是人类唯一的发展道路，自然科学、社会科学、人文精神将交叉融合，世界的文化环境会发生很大的变化，东西方文化将会在激荡过程中进一步融合升华，创造出具有国际化，又有民族特色的新文化。在未来15年，中国要基本完成向一个创新型国家过渡。建立创新体系、创新机制配套的基础是要大幅度提高国民的文化教育水平和科学素质，把我国庞大的人口负担真正转化为无可比拟的创新人力资源。

在中国这样一个大国传播普及科技知识、科学精神是一个宏大的系统工程，需要政府组织倡导和社会各界的积极努力。中国科学院也承担着光荣而艰巨的任务，我们有义务整合全院资源努力把科普工作做大、做好，为国家和社会发挥更大的作用。科学出版社是科普图书出版的一支战略方面军，应该大有作为。《20世纪科普经典特藏》把原汁原味的经典科普大餐奉献给新时代读者，辅之以中文点评是一个很好的尝试。希望这些经典著作能给读者以启发，开拓读者的科学视野，更希望这些经典著作能起到示范的作用，推进我们自己的原创科普和科学文化作品的创作和出版。

路甬祥
2006年七月十七日

点评者序

伽莫夫与斯坦纳德所著的《汤普金斯先生的新世界》一书，可以说是一本科普名著。而且，这本名著本身的演变还有一段很曲折的历史。

先应该简要地介绍一下伽莫夫其人。

伽莫夫（1904～1968），系天才的俄裔美籍科学家，在原子核物理学和宇宙学方面成就斐然，如今在宇宙学中影响最为巨大的大爆炸理论，就有他的重要贡献，甚至于在生物遗传密码概念的提出上，他也是先驱者之一。

除了科学研究之外，科普也是伽莫夫的重要并且极有成就的领域。早年在哥本哈根随量子物理学的一代领袖人物玻尔学习时，他就在玻尔的弟子当中以幽默机智著称。从他的著作中，我们也可以看出其深厚的科学修养和人文修养。他的科普作品数量虽然远远没有像阿西莫夫那样的科普作家那么多，但却本本都有其自身的特色，并且长年拥有大量的读者。从1938年起，他就发表了一系列科学故事，其中成功地塑造了一位名叫汤普金斯的主人公，通过这位主人公的各种经历来传播物理学知识。1940年，他将第一批故事汇集成他的第一部科普著作《汤普金斯先生身历奇境》，1944年，他又将后续故事汇集成《汤普金斯先生探索原子世界》一书。因受读者欢迎，1965年，伽莫夫重新补充新内容，将两本合编为《平装本中的汤普金斯先生》。1968年，伽莫夫去世，在身后这部著作依然是广受欢迎的科普名著，但此后物理学和宇宙学等却在迅速发展，使得书中的内容略显陈旧。在此情况下，剑桥大学出版社大胆邀请了英国著名科普作家斯坦纳德对该书进行全面更新和补写，于是，就成了眼下的这部著作。

伽莫夫在60年代的版本，以及经斯坦纳德修订补充的新版本，都曾有过中译本，均名为《物理世界奇遇记》，分别由科学出版社和湖南教育出版社出版。有人曾说过，上个世纪70年代末科学出版社的中译本，几乎影响了当时国内一代人对物理学的理解和兴趣。不过，后来由湖南教育出版社出版，收入"世界科普名著精选"丛书中的中译本《物理世界奇遇记》（最新版），与这里的英文版在内容结构上又略有不同。

伽莫夫的这本书，可以说是既好读，读起来又有些困难。说好读，是因为伽莫夫的特殊杰出的学识、修养、幽默感和想像力。如果不谈他那些重要的科学贡献，仅就科普著作而言，也足以作为一位兴趣广泛的天才而让人们记住。与其他常见的按主题分类来写作的科普著作不同，伽莫夫完全是以一种大家的写作风格，把数学、物理学的许多内容有机地融合在一起，仿佛作者是想到哪说到哪，将叙述的内容信手拈来，其实，仔细思考，就会感觉到其中各部分内容之间内在的紧密关系。说困难，则是因为按照某种分类，这本书或许可以算作"高级科普"，也就是说，要完全读懂它并不那么容易，需要读者具有某种程度的知识准备，还需要在阅读时随着作者的叙述自己动很多的脑筋来进行思考。

但是，至少有一点可以指出的是，我们也许需要改变一点观念，即读科普书，通常也并不一定非要把书中一切细节都一一彻底搞懂，体会科学家写作的风格和思路、感受科学的思维与美，甚至从中学会一些科学家说话的方式，都可以说是重要的收获。

如今，此书英文（附中文点评）版在国内的出版，可以为学习物理学的学生和物理学爱好者提供一份原汁原味的作品，也可以为英语学习者提供一部很有可读性的优秀科学普及类著作。至于本人在书中的点评，以非常随意的形式写成，有些地方像读书笔记，有些地方像简评，有些地方是感叹，也有些地方则是简要的提示。当然，如果读者略去点评，直接阅读原文，也是可行的阅读方式。

希望读者能够喜欢这本书，并在阅读中有所获益。

刘 兵
2005年10月9日
于清华大学荷清苑

Reviser's Foreword

There cannot be many physicists who have not at one time or other read the Mr Tompkins adventures. Although originally intended for the layperson, Gamow's classic introduction to modern physics has had enduring, universal appeal. I myself have always regarded Mr Tompkins with the greatest affection. I was therefore delighted to be asked to update the book.

如此说来，伽莫夫的这本经典之作可谓"雅俗共赏"。这岂不有些像金庸小说?

A new version was clearly long overdue, so much having happened in the 30 years since the last revision, especially in the fields of cosmology and high energy nuclear physics. But on re-reading the book, it struck me that it was not only the physics that needed attention.

For example, the current output from Hollywood could hardly be regarded as 'infinite romances between popular stars'. Again, ought one to be introducing quantum theory by reference to a tiger shoot, given our modern-day concern for endangered species? And what of 'pouting' Maud, the professor's daughter, 'engulfed in *Vogue*', wanting 'a darling mink coat', and told to 'run along, girlie' at the mere mention of physics. This hardly strikes the right note at a time when strenuous efforts are being made to persuade girls to study physics.

在这里，修订者已经在科普作品中表现出了很好的环保意识和性别意识，可谓是科普观念的与时俱进。

Then there are problems with the plot. While Gamow deserves credit for the innovative way he introduced the physics through a story, the actual storyline has always had its weaknesses. For instance, Mr Tompkins repeatedly learns new physics from his dreams before he has had any chance of being exposed to such ideas (even subliminally) through real life situations involving the professor's lectures or conversations. Or take the case of his holiday at the seaside. He falls asleep in the train and dreams that the professor is accompanying him on his journey. It later turns out that the professor actually is on holiday with him and Mr Tompkins is fearful that he will remember what a fool he made of himself on the train—in his dream?!

At times the physics explanations are not as clear as they might have been. For instance, in dealing with the relativistic loss of simultaneity for events occurring in different locations, a situation is described where observers in two spacecraft are to compare results. But instead of adopting the viewpoint of one of these two frames of reference's, the problem is addressed from a third, and unacknowledged, frame in which both craft are moving. Likewise, the account given of the shooting of the station master, while the porter was apparently reading a paper's at the other end of the platform, does not in fact establish the porter innocence—as is claimed. (The description would need to rule out the possibility of the porter firing the gun before sitting down to read the paper.)

There is the question of what to do with the 'cosmic opera'. The idea of such a work ever being staged at Covent Garden, was, of course, always farfetched. But now we are faced with the added problem that the subject of the opera—the rivalry between the Big Bang theory and the Steady State theory—can hardly be regarded as a live issue today, the experimental evidence having come down heavily in favour of the former. And yet the exclusion of this ingenious, joyful interlude would be a great loss.

Another problem concerns the illustrations. *Mr Tompkins in Paperback* was partly illustrated by John Hookham, and partly by Gamow himself. In order to describe the latest developments in physics, further illustrations would be required, so necessitating yet a third artist. Should one settle for the resulting unsatisfactory clash of styles, or adopt a completely fresh approach?

> 这是一个两难的问题。实际上，伽莫夫的多才多艺，也体现在他的绘画上，他的漫画极有个人风格，有一种质朴但却传神的特征，此书新版中缺少了伽莫夫本人的绘画，实在是一大遗憾！

In the light of these various considerations, a decision had to be made: I could content myself with a minimal rewrite in which I simply patched up the physics and turned a blind eye to all the other weaknesses. Alternatively, I could grasp the nettle and go for a thorough reworking.

I decided on the latter. All the chapters needed work doing on them. Chapters 7, 15, 16 and 17 are entirely new. I decided it would also be helpful to add a glossary. The detailed changes I proposed met with the approval of the Gamow family, the publishers and their panel of advisors —with the notable exception of one consultant who was of the opinion that the text should not in any way be touched. This dissenting view was a signal that I was not going to be able to please everyone! Clearly there will always be those who would rather stay with the original —which is fair enough.

But as far as this version is concerned, it is primarily aimed at those who have yet to make the acquaintance of Mr Tompkins. While trying to remain true to the spirit and approach of Gamow's original, it aims to inspire and meet the needs of the next generation of readers. As such, I would like to think that it is a version George Gamow himself might have written—had he been at work today.

每个时代，科普作品的读者都不相同，但伽莫夫的那种个人魅力，则有着某种超越时代的特征。

Acknowledgements

Thanks are due to Michael Edwards for enlivening the text with his refreshing illustrations. I am grateful to Matt Lilley for his helpful and constructive comments on an early draft. The encouragement and support I received from the Gamow family was much appreciated.

R. STANNARD

Gamow's Preface to *Mr Tompkins in Paperback*

In the winter of 1938 I wrote a short, scientifically fantastic story (not a science fiction story) in which I tried to explain to the layman the basic ideas of the theory of curvature of space and the expanding universe. I decided to do this by exaggerating the actually existing relativistic phenomena to such an extent that they could easily be observed by the hero of the story, C. G. H.* Tompkins, a bank clerk interested in modern science.

I sent the manuscript to *Harper's Magazine* and, like all beginning authors, got it back with a rejection slip. The other half-a-dozen magazines which I tried followed suit. So I put the manuscript in a drawer of my desk and forgot about it. During the summer of the same year, I attended the International Conference of Theoretical Physics, organized by the League of Nations in Warsaw. I was chatting over a glass of excellent Polish miod with my old friend Sir Charles Darwin, the grandson of Charles (*The Origin of Species*) Darwin, and the conversation turned to the popularization of science. I told Darwin about the bad luck I had had along this line, and he said: 'Look, Gamow, when you get back to the United States dig up your manuscript and send it to Dr C. P. Snow, who is the editor of a popular scientific magazine *Discovery* published by the Cambridge University Press.'

So I did just this, and a week later came a telegram from Snow saying: 'Your article will be published in the next

> 斯诺这位"两种文化"问题的提出者,其工作并不仅限于抽象的理论,从这里看,他对"科普作家"伽莫夫的发现亦是大功一件。

---

* The initials of Mr Tompkins originated from three fundamental physical constants: the velocity of light c, the gravitational constant G, and the quantum constant h, which have to be changed by immensely large factors in order to make their effect easily noticeable by the man on the street.

issue. Please send more.' Thus a number of stories on Mr Tompkins, which popularised the theory of relativity and the quantum theory, appeared in subsequent issues of *Discovery*. Soon there after I received a letter from the Cambridge University Press, suggesting that these articles, with a few additional stories to increase the number of pages, should be published in book form. The book, called *Mr Tompkins in Wonderland*, was published by Cambridge University Press in 1940 and since that time has been reprinted sixteen times. This book was followed by the sequel, *Mr Tompkins Explores the Atom*, published in 1944 and by now reprinted nine times. In addition, both books have been translated into practically all European languages (except Russian), and also into Chinese and Hindi.

Recently the Cambridge University Press decided to unite the two original volumes into a single paperback edition, asking me to update the old material and add some more stories treating the advances in physics and related fields which took place after these books were originally published. Thus I had to add the stories on fission and fusion, the steady state universe, and exciting problems concerning elementary particles. This material forms the present book.

A few words must be said about the illustrations. The original articles in *Discovery* and the first original volume were illustrated by Mr John Hookham, who created the facial features of Mr Tompkins. When I wrote the second volume Mr Hookham had retired from work as an illustrator, and I decided to illustrate the book myself, faithfully following Hookham's style. The new illustrations in the present volume are also mine. The verses and songs appearing in this volume are written by my wife Barbara.

G. GAMOW
University of Colorado, Boulder, Colorado, USA

Contents 目录

序·· 路甬祥
点评者序··· 刘兵
Reviser's Foreword····························· 5
Gamow's Preface to *Mr Tompkins in Paperback*
·· 8

1 **City Speed Limit** ······························ 1
  城市速度极限

2 **The Professor's Lecture on Relativity which Caused Mr Tompkins's Dream** ············ 10
  教授那篇使汤普金斯先生进入梦境的相对论演讲

3 **Mr Tompkins Takes a Holiday** ············ 25
  汤普金斯先生请了个疗养假

4 **The Notes of the Professor's Lecture on Curved Space** ······························· 47
  教授那篇关于弯曲空间的演讲稿

5 **Mr Tompkins Visits a Closed Universe** ······· 62
  汤普金斯先生访问一个封闭宇宙

6 **Cosmic Opera** ······························ 73
  宇宙之歌

7 **Black Holes, Heat Death, and Blow Torch** ······ 84
  黑洞、热寂和喷灯

8 **Quantum Snooker** ························ 95
  量子台球

9 **The Quantum Safari** ······················ 118
  量子丛林

10 **Maxwell's Demon** ·················· 128
   麦克斯韦妖
11 **The Merry Tribe of Electrons** ·················· 146
   快乐的电子部落
11½ **The Remainder of the Previous Lecture through which Mr Tompkins Dozed** ··················163
   上一次演讲中汤普金斯先生因为睡着而没有听到的那部分
12 **Inside the Nucleus** ·················· 173
   原子核内部
13 **The Woodcarver** ·················· 184
   老木雕匠
14 **Holes in Nothing** ·················· 196
   虚空中的空穴
15 **Visiting the 'Atom Smasher'** ·················· 205
   参观"原子粉碎机"
16 **The Professor's Last Lecture** ·················· 239
   教授的最后一篇演讲
17 **Epilogue** ·················· 256
   尾声

1 City Speed Limit

城市速度极限

It was a public holiday, and Mr Tompkins, a little clerk of a big city bank, slept late and had a leisurely breakfast. Trying to plan his day, he first thought about going to an afternoon movie. Opening the local newspaper, he turned to the entertainment page. But none of the films appealed to him. He detested the current obsession with sex and violence. As for the rest, it was the usual holiday fare aimed at children. If only there were at least one film with some real adventure, with something unusual and maybe challenging about it. But there was none.

> 作为科普作品，这是一个独特的开场。看来，在伽莫夫通过汤普金斯来表述的审美趣味中，能看得上眼的影片实在是太少了。

Unexpectedly, his eye fell on a little notice in the corner of the page. The town's university was announcing a series of lectures on the problems of modern physics. This afternoon's lecture was to be about Einstein's Theory of Relativity. Well, that might be something! He had often heard the statement that only a dozen people in the world really understood Einstein's theory. Maybe he could become the thirteenth! He decided to go to the lecture; it might be just what he needed.

Arriving at the big university auditorium, he found the lecture had already begun. The room was full of young students. But there was a sprinkling of older people there as well, presumably members of the public like himself. They were listening with keen attention to a tall, white-bearded man standing alongside an overhead projector. He was explaining to his audience the basic ideas of the Theory of Relativity.

> 在今天，我们的科普讲座还会这么吸引人，特别是这么吸引年轻人吗？

Mr Tompkins got as far as understanding that the

whole point of Einstein's theory is that there is a maximum velocity, the velocity of light, which cannot be exceeded by any moving material object. This fact leads to very strange and unusual consequences. For example, when moving close to the velocity of light, measuring rulers contract and clocks slow down. The professor stated, however, that as the velocity of light is 300, 000 kilometres per second (i. e. 186, 000 miles per second), these relativistic effects could hardly be observed for events of ordinary life.

It seemed to Mr Tompkins that this was all contradictory to common sense. He was trying to imagine what these effects would look like, when his head slowly dropped on his chest …

> 科学的理论，与普通人基于日常经验的感受，经常是不一致的！
> 就像电影中常用的切换手法一样，汤普金斯先生已经入梦了！

When he opened his eyes again, he found himself sitting, not on a lecture room bench, but on one of the benches provided by the city for the convenience of passengers waiting for a bus. It was a beautiful old city with medieval college buildings lining the street. He suspected that he must be dreaming, but there was nothing unusual about the scene. The hands of the big clock on the college tower opposite were pointing to five o'clock.

> 以下作者描述的场景，在现实生活中是无法看到的，但它们确实又是对相对论有关速度等理论的形象表达，因此，将这些场景放入主人公的梦境里，便显得合情合理了。

The street was nearly empty—except for a single cyclist coming slowly towards him. As he approached, Mr Tompkins's eyes opened wide with astonishment. The bicycle and the young man on it were unbelievably shortened in the direction of their motion, as if seen through a cylindrical lens. The clock on the tower struck five, and the cyclist, evidently in a hurry, stepped harder on the pedals. Mr Tompkins did not notice that he gained much in speed, but, as a result of his effort, he shortened still further and went down the street looking rather like a flat picture cut out of cardboard. Immediately Mr Tompkins understood what was happening to the cyclist—it was the contraction of moving bodies, about which he had just heard. He felt very pleased with himself. 'Nature's speed limit must be lower here,' he concluded. 'I reckon it can't be much more than 20 m. p. h. They'll not be needing speed cameras in this town.' In fact, a speeding

1 City Speed Limit

ambulance going past at that moment could not do much better than the cyclist; with lights flashing and siren sounding, it was really just crawling along.

Mr Tompkins wanted to chase after the cyclist to ask him how he felt about being flattened. But how was he to catch up with him? It was then he spotted another bicycle standing against the wall of the college. Mr Tompkins thought it probably belonged to a student attending lectures who might not miss it if he were to borrow it for a short while. Making sure no-one was looking, he mounted the bike and sped down the street in pursuit of the other cyclist.

He fully expected that his newly acquired motion would immediately shorten him, and looked forward to this as his increasing girth had lately caused him some anxiety. To his surprise, however, nothing happened;both he and his cycle remained the same size and shape. On the other hand, the scene around him completely changed. The streets grew shorter, the windows of the shops became narrow slits, and the pedestrians were the thinnest people he had ever seen.

'Ah!' exclaimed Mr Tompkins excitedly. 'I get it now. This is where the word *relativity* comes in. Everything that moves relative to me looks shorter for me—whoever works the pedals!'

He was a good cyclist and was doing his best to overtake the young man. But he found that it was not at all easy to get up speed on this bicycle. Although he was working on the pedals as hard as he possibly could,the increase in speed was almost negligible. His legs had already begun to ache, but still he could not manage to pass a lamppost on the corner much faster than when he had just started. It looked as if all his efforts to move faster were leading to nothing. He began to understand now why the ambulance could not do much better than the cyclist. It was then he remembered what the professor had said about the impossibility of exceeding the limiting velocity of light. He noticed, however, that the harder he

> 这是对相对论中一个结论的形象表达,越接近"速度极限"(在相对论中是光速,而在汤普金斯这里的梦中则是某个特定的速度),就要付出越大的努力,需要更多的能量,而且永远也达不到"速度极限"。

1 City Speed Limit

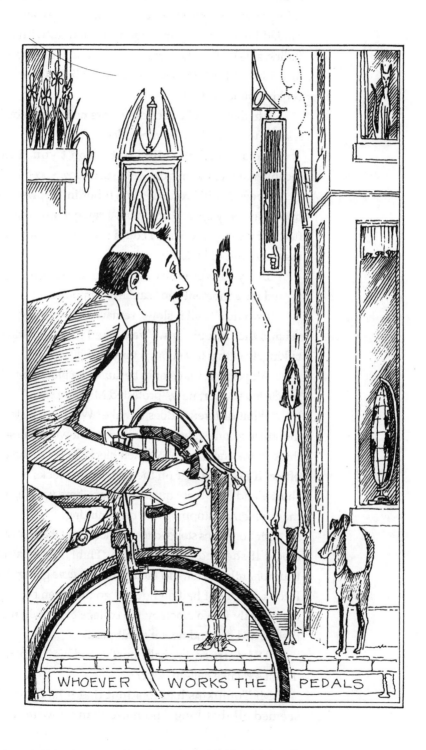

tried, the shorter the city blocks became. The cyclist riding ahead of him did not now look so far away—and indeed he eventually managed to catch up with him. Riding side by side, he glanced across and was surprised to find that both the cyclist and his bike were now looking quite normal.

'Ah, that must be because we are no longer moving relative to each other,' he concluded.

'Excuse me,' he called out, 'Don't you find it inconvenient to live in a city with such a low speed limit?'

'Speed limit?' returned the other in surprise, 'we don't have any speed limit here. I can get anywhere as fast as I wish—or at least I could if I had a motor-cycle instead of this old bike!'

'But you were moving very slowly when you passed me a moment ago,' said Mr Tompkins.

'I wouldn't call it slow,' remarked the young man. 'That's the fifth block we've passed since we started talking. Isn't that fast enough for you?'

'Ah yes, but that's only because the blocks and the streets are so short now,' protested Mr Tompkins.

'What difference does it make? We move faster, or the street becomes shorter—it all comes down to the same thing in the end. I have to go ten blocks to get to the post office. If I step harder on the pedals the blocks become shorter and I get there quicker. In fact, here we are,' said the young man stopping and dismounting.

Mr Tompkins stopped too. He looked at the post office clock; it showed half-past five. 'Hah!' he exclaimed triumphantly. 'What did I tell you. You *were* going slow. It took you all of half an hour to go those ten blocks. It was exactly five o'clock by the college clock when you first passed me, and now it's half-past!'

'Did you *notice* this half hour?' asked his companion. 'Did it *seem* like half-an-hour?'

Mr Tompkins had to admit that it hadn't really seemed all that long—no more than a few minutes.

1 City Speed Limit

Moreover, looking at his wrist watch he saw that it was showing only five minutes past five. 'Oh!' he murmured, 'Are you saying the post office clock is fast?'

'You could say that,' replied the young man. 'Or, of course, it could be your watch running slow. It's been moving relative to those clocks, right? What more do you expect?' He looked at Mr Tompkins with some exasperation. 'What's the matter with you, anyway? You sound like you're from some other planet.' With that, the young man disappeared into the post office.

> 时间的相对性！运动者的时间（由汤普金斯手上的表来表征），与另一个参考系中的时间（邮局的表）是不同的。

Mr Tompkins thought what a pity it was the professor was not at hand to explain these strange happenings to him. The young man was evidently a native, and had been accustomed to this state of things even before he had learned to walk. So Mr Tompkins was forced to explore this strange world by himself. He reset his watch by the time shown on the post office clock, and to make sure it was still going all right, he waited for ten minutes. It now kept the same time as the post office clock, so all seemed to be well.

Resuming his journey down the street, he came to the railway station and decided to check his watch once more, this time by the station clock. To his dismay it was again quite a bit slow.

'Oh dear, relativity again,' concluded Mr Tompkins, 'It must happen every time I move. How inconvenient. Fancy having to reset one's watch whenever you've been anywhere.'

At that moment a well-dressed gentleman emerged from the station exit. He looked to be in his forties. He glanced around and recognised an old lady waiting by the kerb side and went over to greet her. Much to Mr Tompkins's surprise, she addressed the new arrival as 'dear Grandfather'. How was that possible? How could *he* possibly be *her* grandfather?

> 这是对"双生子佯谬"夸张的形象表现。可惜在现实世界中，这种差异小得让人几乎无法觉察。

Overcome with curiosity, Mr Tompkins went up to the pair and diffidently asked, 'Excuse me. Did I hear you rightly? Are you really her grandfather? I'm sorry, but I …'

'Ah, I see,' said the gentleman, smiling, 'perhaps I should explain. My business requires me to travel a great deal.'

Mr Tompkins still looked perplexed, so the stranger continued. 'I spend most of my life on the train. So, naturally I grow old much more slowly than my relatives living in the city. It's always such a pleasure to come back and see my dear little granddaughter. But I'm sorry, you'll have to excuse me, please …' He hailed a taxi, leaving Mr Tompkins alone again with his problems.

A couple of sandwiches from the station buffet somewhat revived him. 'Yes, of course', he mused, sipping his coffee, 'motion slows down time, so that's why he ages less. And all motion is relative—that's what the professor said so that means he will appear younger to his relatives, in the same way as the relatives appear younger to him. Good. That's got that sorted out.'

But then he stopped. He put down the cup. 'Hold on. That's not right,' he thought. 'The grandaughter did *not* seem younger to him; she was older than him. Grey hair is not relative! So what does that mean? All motion is *not* relative?'

He decided to make one last attempt to find out how things really are, and turned to the only other customer in the buffet—a solitary man in railway uniform.

> 由谁对因时间的相对而带来的效应"负责",这是一种出自汤普金斯之口的有趣说法。

'Excuse me,' he began, 'would you be good enough to tell me who is responsible for the fact that the passengers in the train grow old so much more slowly than the people staying at one place?'

'I am responsible for it' said the man, very simply.

'Oh!' exclaimed Mr Tompkins. 'How…'

'I'm a train driver,' answered the man, as though that explained everything.

'A train driver?' repeated Mr Tompkins. 'I always wanted to be a train driver—when I was a boy, that is. But …but what's that got to do with staying young?' he added, looking more and more puzzled.

'Don't know exactly,' said the driver, 'but that's the way it is. Got it from this bloke from the university. Sitting over there we were,' he said nodding at a table by the door. 'Passing the time of day, you know. Told me all about it he did. Way over my head, mind you. Didn't understand a word of it. But he did say it was all down to acceleration and slowing down. I remember that bit. It's not just speed that affects time, he said; it's acceleration too. Every time you get pushed or pulled around on the train— as it comes into stations, or leaves stations—that upsets time for the passengers. Someone who is *not* on the train doesn't feel all those changes. As the train comes into the platform you don't find people standing on the platform having to hold onto rails or what-have-you to stop falling over in the way the passengers on the train do. So that's where the difference comes in. Somehow…' he shrugged.

Suddenly a heavy hand shook Mr Tompkins's shoulder. He found himself sitting not in the station café but on the bench of the auditorium in which he had been listening to the professor's lecture. The lights were dimmed and the room was empty. It was the janitor who had awakened him saying: 'Sorry, sir, but we're closing up. If you want to sleep, you'd be better off at home.' Mr Tompkins sheepishly got to his feet and started towards the exit.

真是个不错的、挺"科学"的梦，醒得也正是时候，否则，就

2 The Professor's Lecture on Relativity which Caused Mr Tompkins's Dream
教授那篇使汤普金斯先生进入梦境的相对论演讲

> 汤普金斯先生梦中的情景虽然生动，但毕竟只是一种生动、形象的比喻，要想更准确地了解相对论的理论内容，恐怕读者就不得不耐下心来"听听"这篇对普通人有些枯燥和困难的教授演讲了。

Ladies and gentlemen:

At a very primitive stage in the development of the human mind there formed definite notions of space and time as the frame in which different events take place. These notions, without essential changes, have been carried forward from generation to generation, and, since the development of the exact sciences, they have been built into the foundations of the mathematical description of the Universe. The great Newton perhaps gave the first clear-cut formulation of the classical notions of space and time, writing in his *Principia*:

> 'Absolute space, in its own nature, without relation to anything external, remains always similar and immovable;' and 'Absolute, true and mathematical time, of itself, and from its own nature, flows equably without relation to anything external.'

So strong was the belief in the absolute correctness of these *classical* ideas about space and time that they have often been held by philosophers as given a priori, and no scientist even thought about the possibility of doubting them.

However, at the start of the present century it became clear that a number of results, obtained by the most refined methods of experimental physics, led to clear contradictions if interpreted in the classical frame of space

2 The Professor's Lecture on Relativity which Caused Mr Tompkins's Dream

and time. This realization brought to one of the greatest twentieth century physicists, Albert Einstein, the revolutionary idea that there are hardly any reasons, except those of tradition, for considering the classical notions concerning space and time as absolutely true, and that they could and should be changed to fit our new and more refined experience. In fact, since the classical notions of space and time were formulated on the basis of human experience in ordinary life, we need not be surprised that the refined methods of observation of today, based on highly developed experimental techniques, indicate that these old notions are too rough and inexact; they have been used in ordinary life and in the earlier stages of the development of physics only because their deviations from the correct notions were too small to be noticeable. Nor need we be surprised that the broadening of the field of exploration of modern science should bring us to regions where these deviations become so very large that the classical notions could not be used at all.

> 对人们平时获得并习以为常的概念进行反思、分析，常常是新发现的开始。

The most important experimental result which led to the fundamental criticism of our classical notions was the discovery that *the velocity of light in a vacuum is a constant (300,000 kilometres per second, or 186,000 miles per second), and represents the upper limit for all possible physical velocities.*

This important and unexpected conclusion was fully supported, for instance, by the experiments of the American physicists Michelson and Morley. At the end of the nineteenth century, they tried to observe the effect of the motion of the Earth on the velocity of light. They had in mind the prevailing view at the time that light was a wave moving in a medium called the aether. As such it was expected to behave in much the same way as water waves move over the surface of a pond. The Earth was expected to be moving through this aether medium in a manner similar to a boat moving over the surface of the water. The ripples caused by the boat appear to a passenger to move away more slowly from the vessel in the direction in which it is travelling than they do to the rear. In one case we have to subtract the speed of the boat from that of the water waves, and in the other we add them. We call this the *theorem of addition of velocities*. This has always been held to be self-evident. In the same way, therefore, one would expect that the speed of light would appear to differ according to its direction relative to the motion of the Earth through the aether. Indeed, it ought to be possible to determine the speed of the Earth with respect to the aether by measuring the speed of light in different directions.

To Michelson and Morley's great surprise, and the surprise of all the scientific world, they found that no such effect exists; the velocity of light was exactly the same in all directions. This odd result prompted the suggestion that perhaps, by an unfortunate coincidence, the Earth in its orbit around the Sun just happened to be stationary relative to the aether at the time the experiment was carried out. To check that this was not so, the experiment was repeated six months later when the Earth was travelling in the reverse direction on the opposite side of its orbit. Again, no difference in the speed of light could be detected.

It having been established that the velocity of light did not behave like that of a wave, the remaining possibility was that it behaved more like that of a projectile. If we were to fire a bullet from a gun in the boat, it would seem to the passenger to leave the moving boat at the same speed in all directions—which is the behaviour Michelson and Morley found for light emitted in all directions from the moving Earth. But in that case, someone standing on the shore would find that a bullet fired in the direction in which the boat was heading would be travelling faster than one fired in the opposite direction. In the first case the speed of the boat would be added to the muzzle speed of the bullet, and in the latter it would be subtracted—again in accordance with the

> 这里描述的速度叠加，正是伽利略时代确立的经典原理，它也与普通人的常识性感觉更为一致。

2 The Professor's Lecture on Relativity which Caused Mr Tompkins's Dream

theorem for the addition of velocities. Accordingly, we would expect that light emitted from a source that was *moving relative to us* would have speeds dependent on the angle of emission to the direction of motion.

Experiment shows, however, that this is also not the case. Take, for example, neutral pions. These are very small sub-atomic particles which undergo decay with the emission of two pulses of light. It is found that these pulses are always emitted with the same speed whatever their direction relative to the motion of the parent pion, even when the pion itself is travelling at a speed close to that of light.

Thus, we find that whereas the first experiment showed that the velocity of light did not behave like that of a conventional wave, this second one shows that it does not behave like a conventional particle either.

In conclusion, we find that the speed of light in a vacuum has a constant value regardless of the movement of the observer (our observations from the moving Earth), or the movement of the source of light (our observations of light emitted from the moving pion).

What of the other property of light I mentioned: it being the ultimate limiting velocity?

'Ah,' you might say, 'but is it not possible to construct a superlight velocity by adding several smaller velocities?'

For example, we could imagine a very fast-moving train with a velocity of, say, three-quarters that of light, and we could have a man running along the roofs of the carriages also with a velocity three-quarters that of light. (I asked you to use your *imagination*!) According to the theorem of the addition of velocities, the total velocity should be $1\frac{1}{2}$ times that of light. That would mean the running man should be able to overtake the beam of light from a signal lamp. It seems, however, that, since the constancy of the velocity of light is an experimental observation, the resulting velocity in our case must be smaller than we expect—the classical theorem for the

与普通人"常识性"感觉不那么一致的"光速不变",是相对论的重要前提,由此出发,才推演出了后来的相对论理论内容。

addition of velocities must be wrong.

The mathematical treatment of the problem-something I do not want to enter into here—leads to a very simple new formula for the calculation of the resulting velocity of two superimposed motions. If v_1 and v_2 are the two velocities to be added, and c is the velocity of light, the resulting velocity comes out to be

$$V = \frac{(v_1+v_2)}{\left(1+\frac{v_1 v_2}{c^2}\right)} \qquad (1)$$

You see from this formula that if both original velocities were small, I mean small as compared with the velocity of light, the second term in the denominator (the bottom bit) of (1) will be so small it can be ignored, giving the classical theorem of addition of velocities. If, however, v_1 and v_2 are not small, the result will always be somewhat smaller than the arithmetical sum. For instance, in the example of our man running along a train, $v_1 = \frac{3}{4}c$ and $v_2 = \frac{3}{4}c$ and our formula gives the resulting velocity $V = \frac{24}{25}c$, which is still smaller than the velocity of light.

You should note that in the particular case when one of the original velocities is c, formula (1) always gives c for the resulting velocity independent of what the second velocity might be. Thus, by overlapping any number of velocities, we can never exceed the velocity of light. This formula has been confirmed experimentally; the addition of two velocities is always somewhat smaller than their arithmetical sum.

Recognizing the existence of the upper-limit velocity we can start on the criticism of the classical ideas of space and time, directing our first blow against the notion of *simultaneity*.

When you say, 'The explosion in the mines near Cape-town happened at exactly the same moment as the ham and eggs were being served in your London apartment,' you think you know what you mean. I am going to show you, however, that you do not. Strictly

> 在相对论中，两个速度相加总会小于这两速度的算术和，这也是与人们的寻常感觉不一致的。

2 The Professor's Lecture on Relativity which Caused Mr Tompkins's Dream

speaking, this statement has no exact meaning.

To see this, consider what method you would use to check whether two events in two different places were simultaneous or not. You would say that the two events were simultaneous if clocks at both places showed the same time. But then the question arises as to how we are to set the distant clocks so that they show the same time simultaneously —and we are back at the original question.

> 在此对人们长期未予置疑的"同时性"的批判性分析中，可操作的验证方法是关键之所在。

Since the independence of the velocity of light in a vacuum on the motion of its source or the system in which it is measured is one of the most exactly established experimental facts, the following method of measuring the distances and setting the clocks correctly on different observational stations should be recognised as the most rational and, as you will agree after thinking more about it, the only reasonable method.

A light signal is sent from station A, and as soon as it is received at station B it is returned back to A. One-half of the time, as read at station A, between the sending and the return of the signal, multiplied by the constant velocity of light, will be defined as the distance between A and B.

> 以下的论证实在有些绕，不耐心琢磨，恐怕就跟不上教授的讲授了。幸好后面还有图解。

The clocks on stations A and B are said to be set correctly if at the moment of arrival of the signal at B the local clock were showing the average of the two times recorded at A at the moments of sending and receiving the signal. Using this method between different observational stations established on a rigid body (in this case, the surface of the Earth) we arrive finally at the desired frame of reference. We can now answer questions concerning the simultaneity of, or the time interval between, two events in different places.

But given that all observers use this method for establishing their frames of reference, will they obtain the same results for their measurements? What if for instance observers are *moving* relative to each other?

To answer this question, suppose that such frames

of reference have been established on two different rigid bodies, say on two long space rockets moving with a constant speed in opposite directions. Let us see how measurements made with these two frames check with one another. Suppose observers are located one at the front, and one at the rear-end, of each rocket. Firstly, each pair of observers needs to set their clocks correctly. This they do using a modification of the above-mentioned method. Using a measuring ruler, they locate the centre of their rocket. Here they place an intermittent source of light. They arrange for the source to emit a pulse of light that spreads outwards towards both ends of the rocket. They agree to set their watches to zero at the instant they receive the pulse from the middle at their respective locations. The light having travelled equal distances to each end, at the same speed, c, our observers have established, according to the previous definition, the criterion of simultaneity in their own system, and have set their watches 'correctly' —from their point of view.

Now they decide to see whether the time readings on their rocket check with those on the other. For example, do the watches of the two observers on rocket 1 show the same time when observed from rocket 2? This can be tested by the following method: At the centre point of each rocket (where the light sources are situated), two electrically charged conductors are installed, in such a way that, when the rockets pass each other and their centres are directly opposite each other, a spark jumps between the conductors. This triggers the two light sources to emit their pulses simultaneously towards the front and rear ends of their respective rockets—as I have shown here in Fig. (a). After a while, according to observers 2A and 2B on rocket 2, we have the situation shown in Fig. (b). Rocket 1 has moved relative to rocket 2. The light beams have moved out equal distances in either direction. But note what has happened. Because observer 1B has moved forward to meet the light beam

2 The Professor's Lecture on Relativity which Caused Mr Tompkins's Dream

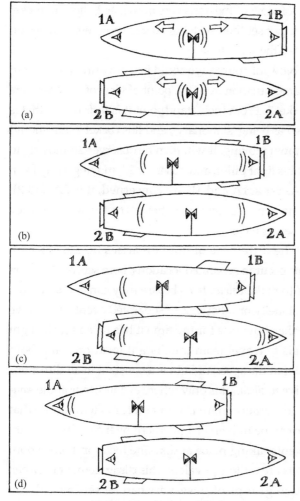

Their watches do not read the same

coming towards her (according, that is, to observers 2A and 2B), the rear-going pulse in rocket I has already reached the position of 1B. According to 2A and 2B, this is because it had less distance to travel. So observer 1B has set her watch going from zero before anyone else! In Fig. (c) the light pulses have reached the ends of rocket 2, and this is when observers 2A and 2B set their watches to zero—simultaneously. It is only when we get to Fig. (d) that the forward-going pulse in rocket 1 catches up with the receding observer 1A—which, according to him is the time to set his watch to zero. Thus, we see that, from the point of view of the observers in rocket 2, those in rocket 1 have not set their watches correctly—their watches do *not* read the same time.

Now, of course, we could just as easily have shown the same situation from the point of view of the observers in rocket 1. From their standpoint it is *their* rocket that is treated as being 'stationary', and it is rocket 2 that should be shown moving. It will then be observer 2B moving to meet his light pulse, and observer 2A moving away from his. As far as 1A and 1B are concerned, it is 2A and 2B who have not set their watches correctly, whereas they themselves have.

The difference of opinion arises because, where events occur in separated locations, both sets of observers have to make *calculations* before they can decide on the simultaneity or otherwise of separated events; they have to make allowance for the time it has taken for the light signals to travel from the distant locations, and both insist that the speed of light is a constant in all directions relative to *them*. (It is only where events occur at the *same* location-where there is no need for calculation—that there can be universal agreement over the simultaneity of events taking place at that one location.) Since both rockets are quite equivalent, this disagreement between the two groups of observers can be settled only by saying that both groups are correct from their own point of view, but the question of who is correct 'absolutely' is one that has no unique answer.

In this way we see that *the notion of absolute simultaneity vanishes, and two events in different places considered as simultaneous from one system of reference will be separated by a definite time interval from the point of view of another system.*

This proposition sounds at first extremely unusual. But let me ask you this: Would it be unusual if I were to say that, having your dinner on a train, you can eat

---

常言"公说公有理，婆说婆有理"，那是因为他们各自在说的时候是"from their own point of view"！

2 The Professor's Lecture on Relativity which Caused Mr Tompkins's Dream

your soup and your dessert in the same point of the dining car, but in widely separated points of the railway track? Of course not. This statement about your dinner in the train can be formulated by saying that *two events happening at different times at the same point in space of one system of reference will be separated by a definite space interval from the point of view of another system.*

I think you will agree that this is a 'trivial' proposition. But now compare it to the previous 'paradoxical' one, and you will see that they are absolutely symmetrical statements. One can be transformed into the other simply by exchanging the words 'time' and 'space'.

Here is the whole point of Einstein's view: Whereas in Newton's classical physics, time was considered as something quite independent of space and motion('flowing equably without relation to anything external'),in the new physics, space and time are closely connected. They represent just two different cross-sections through the one homogeneous 'spacetime continuum' in which all observable events take place. We must not be misled by the very different ways in which we experience and measure the two (one with a ruler, the other with a watch). Physical reality does not consist of a three-dimensional space, together with a separate one-dimensional time. Space and time are indissolubly welded together into a seamless four-dimensional reality—one we refer of as *spacetime*.

> 这样，空间和时间就联系在了一起。

The splitting of this four-dimensional spacetime continuum into a three-dimensional space and a one-dimensional time is purely arbitrary, and depends on the system from which the observations are made. Thus, two events, separated in space by the distance l_1 and in time by the interval t_1 as observed in one system, will be separated by another distance l_2 and another time interval t_2 as seen from another system. It all depends on the particular cross-section one is taking through the four-dimensional reality, and that in its turn depends upon one's motion relative to the events in question.

In a certain sense one can speak about the transformation of space into time, and of time into space. To an extent they can get 'mixed up'. It happens that the transformation of time into space (as in the example of the dinner in a train) is quite a common notion for us. On the other hand, the transformation of space into time, resulting in the relativity of simultaneity, seems unusual. The reason for this is that if we measure distances in, say, 'metres', the corresponding unit of time should not be the conventional 'second', but a more rational unit of time representing the interval of time necessary for a light signal to cover a distance of one metre, i. e. 0. 000,000,003 second. If we were naturally sensitive to time intervals of that kind of duration, the loss of simultaneity would have always been manifestly obvious to us. It is the fact that, in the sphere of our ordinary experience, the transformation of space intervals into time intervals leads to differences in observation that are practically unobservable, which has led to the classical view of time as being something absolutely independent and unchangeable.

> 对人生而言如此短暂的时间，在科学的分析中，却至关重要！

When investigating motions with very high velocities, however, such as those encountered when electrons are thrown out from radioactive atomic nuclei—where the distances covered in a certain interval of time are of the same order of magnitude as the time expressed in rational units—then one necessarily meets with the effects we have discussed, and the theory of relativity becomes of great importance. Even in the region of comparatively small velocities, as, for example, the motion of planets in our solar system, relativistic effects can be observed. This is due to the extreme precision of astronomical measurements. Such observation of relati-vistic effects requires measurements of changes of planetary motion amounting to a fraction of an angular second per year.

So, as I have tried to explain to you, our examination

2 The Professor's Lecture on Relativity which Caused Mr Tompkins's Dream

of the notions of space and time leads us to the conclusion that space intervals can be partially converted into time intervals, and vice versa. This means that the numerical value of a given distance or a period of time can be different when measured from different moving systems.

A comparatively simple mathematical analysis of this problem, into which I do not, however, want to enter in these lectures, leads to a definite formula for the change of these values. For those of you interested, it works out that any object of length l_0, when moving relative to an observer with velocity v, will appear to be shortened by an amount depending on its velocity. Its measured length, l, will be

$$l = l_0 \sqrt{\left(1 - \frac{v^2}{c^2}\right)} \qquad (2)$$

From this you will see that as v becomes very close to c, l becomes smaller and smaller. This is the famous relativistic length contraction. I hasten to add that this is the length of the object in the direction of motion. Its dimensions at right angles to that direction remain unaltered. The object in effect becomes flattened in the direction of motion.

所以汤普金斯梦中的骑车人会"变瘦",而不是"变胖"、"变高"或"变矮"。

Analogously, any process taking time t_0 will be observed from a system moving with velocity v relative to that process to be taking a longer time t, given by

$$t = \frac{t_0}{\sqrt{\left(1 - \frac{v^2}{c^2}\right)}} \qquad (3)$$

Note that as v increases, so does t. Indeed, as v approaches the value of c, t becomes so large that the process essentially comes to a halt. This is known as relativistic *time dilation*. It is the origin of the idea that if one were to have astronauts travelling close to the speed of light, their ageing processes would slow down so much they would effectively get no older—they could live forever!

Don't forget that these effects are absolutely symmetrical as between frames of reference in uniform relative motion. Whereas people standing on the station platform will consider that passengers on a fast-moving train are very thin and move about the train very slowly, with watches on their wrists that are going slow, the passengers on that train will think the same about the people they see outside standing on the platform; the station will be squashed up and everything happening there will be in slow motion.

At first sight this might strike you as paradoxical. Indeed, the problem has become known as the 'twin paradox'. The idea is that you have two twins, one of whom goes on a journey, leaving the other at home. According to the theory I have presented, each twin will believe it is the other who is ageing less quickly, based on their observations of the other and the calculations they have had to make as regards how long the light signals have taken to reach them. The question is what will they discover when the travelling twin returns and a *direct comparison* can be made between them—a comparison that no longer requires any calculations to be made because they are once more in the same location? (Obviously they can't *both* be older than the other.) The resolution of the problem comes from the recognition that the two twins are *not* on the same footing. In order for the travelling twin to return, she must undergo acceleration—first of all slowing down, and then reaccelerating in the opposite direction. Unlike her twin brother, she has not remained in a state of uniform motion. Only the stay-at-home twin has abided by this condition, and so it is this twin who finds himself vindicated in his belief that his sister is now younger than himself.

One more point before I end. You might be wondering what prevents us accelerating an object to a speed greater than that of light. Surely, you might be thinking,

> 比喻是这么说，不过在科学实验的验证中，实验对象可不是人类的双生子（twin），而是物质粒子。

2 The Professor's Lecture on Relativity which Caused Mr Tompkins's Dream

if I push hard enough and for long enough on the object so that it is always accelerating, eventually it must reach any desired speed.

According to the general foundation of mechanics, the mass of a body determines the difficulty of setting it into motion or accelerating the motion already existing; the larger the mass, the more difficult it is to increase the velocity by a given amount. The fact that no object under any circumstances can exceed the velocity of light leads us to a possible interpretation of what is going on. This holds that the increased resistance to further acceleration is due to an increase in the object's mass. In other words, its mass must increase without limit when its velocity approaches the velocity of light. Mathematical analysis leads to a formula for this dependence, which is analogous to the formulae (2) and (3). If m_0 is the mass for very small velocities, the mass m at the velocity v is given by

> 奥运的"更快"目标,在这里受到了遏制,遇到了极限。物理学家还给出了为什么无法超越极限的理由。

$$m = \frac{m_0}{\sqrt{\left(1-\frac{v^2}{v^2}\right)}} \qquad (4)$$

From this we see that the resistance to further acceleration becomes infinite when v approaches c—hence c is the ultimate speed. A good demonstration of the relativistic change of mass can be observed experimentally on very fast-moving particles. Take, for example, electrons. These are the tiny particles found within atoms, moving about the atom's central nucleus. They are easy to accelerate because they are so light. When electrons are stripped out of their atoms and subjected to powerful electric forces in special particle accelerators, they can be made to reach speeds that are within a tiny fraction of the speed of light. At such speeds their resistance to further acceleration can be the equivalent of a particle of mass 40,000 times greater than the normal mass of the electron—as has been demonstrated at a laboratory at Stanford in California.

> 像这样的"科学演讲"的文字,许多科学家、科普作家都会写,伽莫夫的与众不同,在于他只是将此作为"插曲"。接下来,读者就又该见到那位可爱又略微有些可笑的汤普金斯先生了。

Not only that, but time dilation has also been demonstrated. In the high energy physics laboratory called CERN, just outside Geneva in Switzerland, unstable muons (a type of fundamental particle that normally undergoes radioactive disintegration after about one millionth of a second) have been found to live longer by a factor of thirty when travelling at high speed around a circular machine shaped like a large hollow doughnut. At the speed the muons were travelling, a factor of thirty is exactly the value expected on the basis of the above formula for time dilation.

Thus, for such velocities, the classical mechanical approximations become absolutely inadequate, and we enter into a domain where the application of the theory of relativity becomes inescapable.

3 Mr Tompkins Takes a Holiday
汤普金斯先生请了个疗养假

Several days after that first lecture, Mr Tompkins was still intrigued by his dream concerning the relativistic city. He was particularly puzzled over the mystery of how the train driver had been able to prevent the passengers from getting old. Each night he went to bed with the hope that he would see this interesting city again. But it was not to be. Being a somewhat timid and anxious man, his dreams were mostly unpleasant. Last time it was the manager of the bank who was firing him for being slow preparing his accounts. Mr Tompkins's attempted excuse based on relativistic time dilation had fooled no-one. He decided he needed a holiday. Thus, he found himself sitting in a train, watching through the window the grey roofs of the city suburb gradually giving place to green country meadows as he headed for a week-long stay by the sea. He had unfortunately had to miss the second lecture in the series, but had managed to get hold of a photocopy of the professor's notes from the departmental secretary. He had already tried to make sense of them, but had not got far. Having brought them along with him, he pulled them out of the suitcase and began studying them once more. Meanwhile, the railway carriage rocked him pleasantly…

When he lowered the notes and looked out of the window again, the landscape had changed considerably. The telegraph poles were so close to each other they looked like a hedge, and the trees had extremely narrow crowns, rather like Italian cypresses. To add to his

又一个梦开始了！

delight, who should be sitting opposite him but the professor! He must have boarded while Mr Tompkins had been busy reading.

Plucking up courage, Mr Tompkins decided to take advantage of the occasion.

'I take it we're in the land of relativity,' he remarked.

'Yes indeed,' replied the professor, 'you're familiar with it…?'

'I was here once before.'

'You're a physicist—an expert on relativity?' enquired the professor.

'Oh no,' protested Mr Tompkins in some confusion. lecture so far.'

'Good. Never too late. Fascinating subject. Where exactly are you studying?'

'At the university. It was *your* lecture I attended.'

'Mine?!' exclaimed his companion. He looked hard at Mr Tompkins, then flashed a smile of recognition. 'Ah yes. The man who crept into the back late! I remember now. Yes, I thought your face was familiar.'

'I hope I didn't disturb…' mumbled Mr Tompkins apologetically. He desperately hoped the observant professor had not noticed he had eventually dozed off in his lecture.

'No, no. That's all right,' was the reply. 'Happens all the time.'

Mr Tompkins reflected for a moment, then ventured, 'I don't want to impose on you, but I was wondering if I might ask you a question—just a short one? Last time I was here, I met a train driver who insisted that the reason why passengers grow old less quickly than the people in the city—and not the reverse—was all to do with the fact that the train stops and starts. I didn't understand…,

The professor looked thoughtful, and then began:

'If two people are in uniform relative motion, then each will conlude that the other is ageing less quickly

> 以平常人可见的方式观察日常生活中的相对论效应，恐怕也只能在梦境中了。

3 Mr Tompkins Takes a Holiday

than themselves—relativistic time dilation. A passenger on the train will think that the booking clerk in the station is ageing less quickly than she is; likewise, the booking clerk will conclude that it is she who is ageing less quickly than he.'

'But they can't both be right,' objected Mr Tompkins.

'Why not? They are both right—from their own point of view.'

'Yes, but who is *really* right?' Mr Tompkins insisted.

'You can't ask general questions like that. In relativity, your observations must always be with respect to a particular observer—an observer with a well-defined motion relative to whatever is being observed.'

'But we know it's the passenger who ages less than the clerk—it is not the other way round.' Mr Tompkins went on to describe his encounter with the much travelled gentleman and his granddaughter.

'Yes. yes,' interrupted the professor somewhat impatiently. 'It's the twin paradox all over again. I dealt with that in my first lecture if you recall. The grandfather is subject to acceleration; unlike the granddaughter, he does *not* remain in a state of uniform constant motion. So she is the one who correctly expects her grandfather to have aged less when he gets back and they can compare themselves side by side.'

'Yes. I see that,' agreed Mr Tompkins. 'But I still don't get it. The granddaughter can use the time dilation of relativity to understand why her grandfather has aged less; that's not a problem. But won't the grandfather be at a loss to understand how his granddaughter has aged *more*? How does he account for *that*?'

汤普金斯一时搞不懂，对许多普通读者来说，恐怕也有关似的感觉。

'Ah,' replied the professor, 'but that's what I was dealing with in the second lecture, remember?'

It was here Mr Tompkins had to explain how he had missed it—but was nevertheless trying to catch up by reading the notes.

'I see,' resumed the professor, 'Well, let me put

it like this: In order for the *grandfather* to understand what's going on, he must take account of what he reckons is happening to his granddaughter while he *changes* his motion.'

'And what would that be?' enquired Mr Tompkins.

'Well, while he is travelling along with uniform velocity, his granddaughter ages less—the usual time dilation. But once the driver applies the brakes, or later accelerates back on the return journey, then that has precisely the opposite effect on her ageing processes; they appear to the grandfather to *speed up*. It's during those brief spells of *non* uniform motion that her ageing races way ahead of the grandfather's. So, even though she then resumes her normal slower ageing rate during the uniform coasting home, the net effect when he gets back is that he expects her to have aged more than heand that is what he finds.'

> 这里的问题开始复杂化，引入了速度的变化。而"加速度"的因素，已经是广义相对论的问题了。

'How extraordinary,' observed Mr Tompkins. 'But do scientists have any proof of this? Are there any experiments that show this differential ageing?'

'Certainly. In my first lecture I mentioned the unstable muons circulating around that hollow doughnut at the CERN laboratory in Geneva. Because they had a speed close to that of light, they took thirty times longer to disintegrate than muons that just sit still in the laboratory. The moving muons are like the grandfather; they are the ones performing the round trip journey and experiencing all the forces needed to steer them on their course and bring them back to their starting point. The stationary muons are like the granddaughter; they age at the normal rate; they disintegrate or 'die' quicker than the moving ones.

> 这种利用微观粒子的实验，才是科学家真正用来检验相对论效应的方法，而此书中汤普金斯梦境中的场景以及其中表现出来的相对论效应，只是极度夸张了的比喻而已。

'In fact there is another way of checking this out—an indirect one: The conditions existing in a non-uniformly moving system are analogous, or should I say identical, to the result of the action of a very large force of gravity. You may have noticed that when you are in a lift which is rapidly accelerated upwards it seems

> 据说相对论的提出者爱因斯坦，曾真的在头脑中想像过这种电梯中的场景。

3 Mr Tompkins Takes a Holiday

that you have grown heavier; on the contrary, if the lift starts downward (you realise it best when the cable breaks) you feel as though you were losing weight. The explanation is that the 'gravitational field' created by acceleration is added to, or subtracted from, the gravity of the Earth. This equivalence between acceleration and gravity means that we can investigate the effect of acceleration on time by noting what effect gravity has. It is found that the Earth's gravity causes atomic vibrations to occur faster at the top of a tall tower than they do at the bottom. And this is exactly what Einstein predicted would be the effect of acceleration.'

> 在这里，又引入了引力的概念，以及引力与加速度在引起效应上的等效的重要概念。

Mr Tompkins frowned. He did not see the connection between speeded up atomic vibrations at the top of a tower, and the granddaughter's supposed speeded up ageing. Noting his puzzlement, the professor continued.

'Suppose you are at the bottom of the tower looking up at those speeded up atomic vibrations occurring at the top. You are being acted on by an external force: the floor is pushing up on you to counter gravity. It is the fact that this upward force has come into play that increases the time processes of anything lying in the upwards direction. The further away from you the atoms are, the greater will be what we call *the gravitational potential difference* between you and those atoms. That in turn means the more speeded up those atoms will be compared with atoms you have with you at the bottom of the tower.

'Now, in the same way, if you are acted upon by an external force in this train...' He paused. 'In fact, I do believe we *are* slowing down; the driver has applied the brakes. Excellent. At this very moment the back of your seat is applying a force to you altering your velocity. It is acting in a direction towards the back of the train. While this is going on, the time processes of everything occurring down the line in that direction will be speeded up. And if that's where your granddaughter is, that's what will be happening to her.'

'Where are we anyway?' he asked peering out of the window.

The train was passing slowly through a little countryside station. There was no-one on the platform apart from a ticket collector and, at the other end of the platform, a young man sitting at the window of the booking office, reading a newspaper. Suddenly the ticket collector threw his hands into the air and fell down on his face. Mr Tompkins did not hear the sound of shooting, which was probably lost in the noise of the train, but the pool of blood forming round the body of the collector left no doubt as to what had happened. The professor immediately pulled the emergency cord and the train stopped with a jerk. When they got out of the carriage the young booking clerk was running towards the body, carrying a gun. At that moment a policeman came on the scene.

> 这里，真像警匪片中的情节！

'Shot through the heart,' said the policeman after inspecting the body. He turned to the young man. 'I am arresting you for the murder of the ticket collector. Hand over that gun.'

The clerk looked in horror at the gun.

'It's not mine!' he cried. 'I just picked it up. It was lying over there. I was reading and heard this shot and came running. And there was the gun lying on the ground. The murderer must have thrown it down as he made his getaway.'

> 把枪捡起来拿在手中就是个错误。

'A likely story,' said the policeman.

'I tell you,' insisted the young man, ' I never killed him. Why would l want to do a thing like that to the old boy...?'

He looked around desperately. 'You ,' he said pointing to Mr Tompkins and the professor. 'You must have seen what happened. These gentlemen can testify that I am innocent.'

3 Mr Tompkins Takes a Holiday

> 看看，这位警察居然懂得物理学中关于同时性和参考系（即文中的 the system from which you observe）的概念，不过，在这样一个相对论效应如此明显的国度中当警察，对相对论的知识的掌握到也真是不可缺少的基本从业要求。

'Yes,' confirmed Mr Tompkins,' I saw it all. This man was reading his paper when the ticket collector was shot. He did not have the gun with him at the time.'

'Huh! But you were on the train,' said the policeman dismissively.'You were moving weren't you. *Moving*! What you saw counts for nothing. That's no evidence at all. As seen from the *platform* the man could have taken out the gun and shot the victim, even thougt at the time of the death it seemed to *you* on the train that he was still reading. Simultaneity depends on the system from which you observe it, right? I know you mean well, sir, but you're just wasting my time. Come along with me, 'he said, turning to the unfortunate clerk.

'Er, excuse me, officer,' interrupted the professor, 'but I think you are making a mistake—a *serious* mistake. It's true, of course, that the notion of simultaneity is highly relative in your country. It is also true that two events in different places could be simultaneous or not, depending on the motion of the observer. But, even in your country, no observer can see the consequence before the cause. (I take it you've never received a letter before it was sent, or got drunk before opening the bottle?) Now the fact is that we saw the young man take hold of the gun *after* the ticket collector fell dead. As I understand it, you are supposing that because of the motion of the train, we could have seen the collector get shot before his murderer fired the gun that caused his death. Respectfully, I would point out to you that this is an impossibility—even in your country. I know that in your police force you are taught to work strictly by what is written in your instruction manual. I suspect if you look, you'll probably find something about this…'

> 这里涉及到另一个更基本的规则：因果性。它要比相对论更根本，在科学中是不可以违反的。

The professor's authoritative tone made quite an impression on the policeman. Pulling out his pocket book of instructions, he thumbed through it slowly. Soon a sheepish smile of embarrassment spread across his big, red face.

3 Mr Tompkins Takes a Holiday

'Yes, I think I see what you're on about, sir,' he admitted. 'Here it is: section 37, subsection 12, paragraph e. "If a reliable observation is made from any moving system whatsoever, that the suspect was at a distance d from the scene of the crime within a time interval $\pm cd$ of the instant at which the crime was committed (c being the natural speed limit), then the suspect could not have been the cause of the crime and thus has an acceptable alibi."'

'I am very sorry, sir,' he mumbled to the clerk. 'There seems to have been some mistake. I do apologise.'

The young man looked relieved.

Turning to the professor, the policeman added, 'And thank you very much, sir. I'm new to the force, you see. I'm still having to get the hang of all these rules. I must say you've saved me from a lot of trouble back at headquarters. But if you'll excuse me now, I must report the murder.'

With that he began speaking into his mobile radio. A minute later, just as Mr Tompkins and the professor were reboarding the train, having taken their leave of the grateful clerk, the officer called out to them. 'Good news! They appear to have caught the real murderer. My colleagues have picked up a suspect running away from the station. Thank you once more!'

Having resumed their seats, Mr Tompkins asked, 'I may be stupid, but I still don't feel I have fully grasped all this business about simultaneity. Am I right in saying that it really has no meaning in this country?'

'It has,' was the answer, 'but only to a certain extent; otherwise I should not have been able to help the clerk just then. You see, the existence of a natural speed limit for the motion of any object, or the sending of any signal, makes simultaneity in our ordinary sense of the word lose its meaning. Let me put it this way. Suppose you have a friend living in a far-away country with whom you correspond by air mail. Let's say it takes three days for a letter to make the journey. Something

> 瞧，这里的法律条款都离不开相对论的内容！

> 物体或信号传播有速度极限，这是相对论的一个前提，在我们的世界中，这个速度上限就是光速。

33

happens to you on Sunday and you learn that the same thing is going to happen to your friend. It is clear that you cannot let him know about it before Wednesday. On the other hand, if he knew in advance about the thing that was going to happen to you, the last date to let you know about it would have been the previous Thursday. Thus for three days beforehand your friend was not able to influence your fate on Sunday, nor for three days afterwards could he in turn be affected by what happened to you on that Sunday. From the point of view of causality he was, so to speak, excommunicated from you.'

'What about sending a message via email?' suggested Mr Tompkins.

> email? 伽莫夫的时代哪有什么电子邮件!

'I was assuming—for the sake of argument-that the velocity of the plane carrying the mail was the maximum possible velocity. In point of fact, it is the velocity of light(or any other form of electro-magnetic radiation—such as radio waves) that is the maximum velocity. You cannot send a signal, or have any causal influence, faster than that.'

'I'm sorry, you've lost me,' said Mr Tompkins. 'What has all this to do with simultaneity?'

'Well,' replied the professor. ' Take Sunday lunch, for instance. Both you and your friend have Sunday lunch. But do you have it at the same time—simultaneously? One observer might say yes. But there would be others, making their observations from different trains, say, who would insist that you ate your Sunday dinner at the same time as your friend had his Friday breakfast or Tuesday lunch. But—and this is the point—in no way could anybody observe you and your friend simultaneously having meals more than three days apart. If you did, you'd get into all sorts of contradictions. For example, it would be possible for you to send by mail train your Sunday lunch leftovers for your friend to eat for his Sunday lunch. How could an observer then conclude that you were eating your Sunday lunches simultaneously if you had clearly

3 Mr Tompkins Takes a Holiday

already finished yours? And another thing...'

At this point their conversation was interrupted. A sudden jolt awoke Mr Tompkins. The train had come to a halt at its destination. Mr Tompkins hurriedly gathered up his things, stepped down from the train, and went in search of his hotel.

梦醒了。

      🙰   🙰   🙰

Next morning, when Mr Tompkins came down to have his breakfast in the long glass verandah of the hotel, a surprise awaited him. At the table in the opposite corner sat the professor! This was not actually as great a coincidence as one might think. At the time Mr Tompkins had gone to the university to collect the lecture notes, the secretary had drawn his attention to a notice stating that the following week's lecture had been cancelled. Mr Tompkins learned from the secretary that this was because the professor was taking a week's vacation. On remarking to her that he hoped he was going somewhere nice, she had mentioned the name of the resort. It was one of Mr Tompkins's favourites, though he had not been there for a number of years. It was this that had given him the idea of following the professor's, example. That was how they came to end up in the same seaside town—though it was an added bonus for Mr Tompkins to find himself by chance at the very same hotel as the professor.

But what took Mr Tompkins's eye even more than the professor was the person to whom he was talking: a casually dressed woman, not exactly beautiful but certainly striking to look at, shortish but elegant, with long hands which she moved expressively as she spoke and laughed. Mr Tompkins reckoned she must be in her early 30s—probably a few years younger than himself. He wondered what a young woman like that saw in an old man like the professor.

哈哈！警匪片之后，该上演言情片了。

At that moment she happened to glance in his direction. To his embarrassment, before he could look away, she caught him staring at her. She gave him

Pleased to meet you, Maud

a polite, little smile, before immediately turning back to her companion. The professor had meanwhile followed her gaze, and was now examining him intently. As their eyes met he gave a brief, quizzical nod as if to say 'Don't I know you from somewhere?'

Mr Tompkins felt he had better go across and introduce himself. It felt odd doing it a second time, but he realized, of course, that yesterday's encounter had been but a dream. The professor warmly invited him to change tables and come and join them.

'This, by the way, is my daughter, Maud,' he said.

'Your daughter!' exclaimed Mr Tompkins. 'Oh.'

'Is something wrong?' enquired the professor.

'No, no,' stammered Mr Tompkins. 'No. Of course not. Pleased to meet you, Maud.'

She smiled and offered her hand. After they had resumed their seats and ordered breakfast, the professor turned to Mr Tompkins and asked 'So, what did you make of all that stuff on curved space—in the last lecture...?'

'Dad!' Maud gently admonished him. But he ignored her. Again, for what seemed the second time, Mr Tompkins had to apologize for having missed it. The professor was,

一场罗曼史由此开场。

however, impressed that he had gone to the trouble of procuring the lecture notes and was trying to catch up.

'Good. You're obviously keen,' he said. 'If we get bored with all this lying around doing nothing all day long, I could give you a tutorial.'

'Dad!' exploded Maud indignantly.'That's not what we're here for. You're supposed to be getting away from it all for a week.'

He just laughed. 'Always telling me off,' he said, patting the back of her hand affectionately. 'The holiday's her idea.'

'And your doctor's, remember.' she reminded him.

'Well, anyway,' said Mr Tompkins, 'I certainly got a lot out of the first lecture.' He laughed as he went on to describe his dreams about relativity land—how the streets had become visibly shrunk, and how the time dilation effects had been greatly magnified.

'Now that's what I've been telling you about,' said Maud to her father. 'If you're to give lectures to the public, you simply *have* to make them more concrete. People have to relate the effects you're talking about to everyday life. I reckon you ought to include relativity land in your lectures; take a tip from Mr Tompkins here. You're too abstract—too, too...*academic.*'

这几乎就是对科普方式的当代诠释了。

'Too academic,' the professor repeated with a chuckle. 'She's always on about that.'

'Well, you *are.*'

'OK, OK,' he conceded. 'I'll think about it. Mind you,' he added, 'it's not *right*. Even if the speed limit were something like 20 miles per hour, you wouldn't *see* a passing bicycle shortened.'

'You wouldn't?' queried Mr Tompkins, looking confused.

'Not as such. No. The point is that what you see—with your eyes, or what you would photograph with a camera—depends on what light arrives at the eye or lens *at the same instant in time.* Now, if light from the rear of the bike has further to travel to you than light from the front, then the light arriving at a particular point in time from each end must have started out at different times—when the bike was in different positions. Light from the rear must have started out from—and will appear to be coming from—the place where the rear of the bike was when it was further down the road…'

Mr Tompkins wasn't quite following this, so the professor stopped. He thought for a moment, then shrugged.

'It's a small point. It's just that the finite speed of light *distorts* what you see. What you would actually *see* in relativity land is a bike that appeared to be *rotated*.'

'*Rotated*!' exclaimed Mr Tompkins.

'Yes. That's how it happens to work out. It appears rotated, rather than shortened. It's only when you take this raw observation—the data on your photograph, say—and you make due allowance for the different journey times of the light arriving at different points on the photo, that you calculate (note, *calculate* rather than see)—it's only then you conclude that to get that picture, the bike must be length-contracted.'

'There you go again. Academic nit-picking,' interrupted Maud.

'*Nit-picking*!' the professor exploded.' It's nothing of the sort…'

'I have to go back to my room. I need my sketch pad,' she announced. 'I'll leave you two to it. See you for lunch.'

Maud having left, Mr Tompkins remarked, 'I take it she likes doing a bit of drawing then?'

'A bit of…' The professor gave him a warning look. 'I shouldn't let her hear you say that. Maud is an artist—a professional artist. She's made quite a name for herself. It's not everyone gets a retrospective exhibition in a Bond Street gallery. And there was that profile on her in *The Times* last month.'

> 这倒是艺术与科学之分离的一个例子。

3 Mr Tompkins Takes a Holiday

'Really!' exclaimed Mr Tompkins. 'You must be very proud of her.'

'I am indeed. It's all turned out well, *very* well—in the end.'

'In the end? What do you mean...?'

'Oh nothing. It's just that this wasn't exactly what I had had in mind for her. She was cut out to be a physicist at one stage. Very good she was—top of her year in both maths and physics at college. Then suddenly, she gave it all up. Just like that ...' His voice trailed off.

Pulling himself together, he continued 'But as I said, she's made a success of herself—and she's happy. What more could I want?' He glanced out of the dining room window. 'Care to join me? We could grab a couple of deck chairs before they all go, and...' he added conspiratorially, making sure Maud was not around, 'we could talk shop.'

They made their way to the beach and settled down in a quiet spot.

'So,' began the professor, 'let's think about curved space. We can do this best by thinking of a surface—a two-

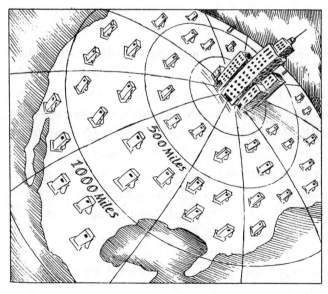

Oil stations concentrated near Kansas Cify

dimensional surface—like that of the Earth. Imagine some oil tycoon decides to see whether his petrol stations are distributed uniformly throughout some country, say America. To do this, he gives orders to his office, somewhere in the middle of the country (Kansas City, say). They are to count the number of stations within a certain distance of the city, then the number within twice that distance, three times, and so on. He remembers from his school days that the area of a circle is proportional to the square of its radius, and expects that in the case of a uniform distribution, the number of stations counted should increase like the sequence of numbers 1, 4, 9, 16, and so on. When the report comes in, he is surprised to see that the actual number of stations is increasing somewhat more slowly, going, let us say,1, 3. 9, 8. 6 , 14. 7, and so on. "I don't understand," he would exclaim; "my managers do not seem to know their job. What is the great idea of concentrating the stations close to Kansas City?" But is he right in this conclusion?'

'It sounds like it,' agreed Mr Tompkins.

'He is not.' declared the professor. 'He has forgotten that the Earth's surface is not a plane but a sphere. And on a sphere the area within a given radius grows more slowly with the radius than on a plane. Take that ball over there,' he said, indicating towards a girl throwing a beach ball to her father. 'Suppose that were to be a globe with a north pole marked on it. If you start from that north pole as centre, then the circle with radius equal to a half meridian is the equator, and the area included is the northern hemisphere. Increase the radius twice and you will get in all the Earth's surface: the area will increase only twice instead of four times as it would on a plane. The difference is due to the positive curvature of the surface. OK?'

'Yes, I think so,' said Mr Tompkins. 'But why did you say "positive"? Is there such a thing as "negative curvature"?'

> 为什么球面与平面不同?
> 由此引入了正曲率和负曲率的重要概念。

3 Mr Tompkins Takes a Holiday

'Certainly.' His eyes roved around the beach searchingly. 'There! That's an example of it right over there,' he said, pointing to a donkey giving a ride to a boy. 'The saddle. The surface of that donkey's saddle is an example of negative curvature.'

'A saddle?' repeated Mr Tompkins.

'Yes, or on the surface of the Earth, a saddle pass between two mountains. Suppose a botanist lives in a mountain hut situated on such a saddle pass and is interested in the density of growth of pines around the hut. If he counts the number of pines growing within one hundred, two hundred, and so on metres from the hut, he will find that the number of pines increases *faster* than the square of the distance—the opposite of what we had for the globe. For a saddle surface, the area included within a given radius is larger than on a plane. Such surfaces are said to possess a negative curvature. If you try to spread a saddle surface on a plane you will have to make folds in

> 这里又是借助地球上的地形作为例子来讲解曲率。

A mountain hut in a saddle pass

it, whereas doing the same with a spherical surface you will probably tear it if it is not elastic.'

'I see,' said Mr Tompkins.

'Another thing about these saddle surfaces,' the professor continued. 'The area of a sphere is finite ($4\pi r^2$); the surface closes back on itself. But it's not like that with a saddle. A saddle surface could, in principle, be extended indefinitely in all directions. It's an "open" surface, not a "closed" one. Of course, in my example of a saddle pass the surface ceases to possess negative curvature as soon as you walk out of the mountains and go over into the positively curved surface of the Earth. But of course you can imagine a surface which preserves its negative curvature everywhere.'

'OK,' said Mr Tompkins. 'But if you'll forgive me, this all seems very straightforward. Why are you telling me...?'

'Ah, well, the point is that exactly the same kind of thinking applies to THREE-dimensional space—not just to the two-dimensional "spaces" or surfaces we've been dealing with so far. Three-dimensional space can be curved.'

'But how...?'

'Same reasoning as before. We use the same tech-nique. Let's suppose we have objects distributed uniformly throughout space—three-dimensional space now, not just petrol stations distributed on the two-dimensional surface of the Earth. They might be stars now—or better still galaxies (great swirling collections of stars scattered throughout space), orclusters of galaxies. Suppose the clusters were to be more-or-less uniformly distributed—meaning the distance between them was always the same. OK, you count their number within different distances from you. If this number grows as the cube of the distance, the space is flat. (You know of course that the volume of a sphere goes up as the cube of its radius—according to normal Euclidean geometry?)'

> 讨论由二维的"面"，推广到"三维"的空间，还是在讲曲率问题，实际上这里讲空间的正、负曲率才是正题，前面讲"面"只是准备和过渡，讲空间的曲率又是为后面讲宇宙做准备。

3 Mr Tompkins Takes a Holiday

Mr Tompkins nodded.

'OK, then,' the professor continued. 'If that's how the number of galaxy clusters goes up, then the space is said to be 'flat'—it's genuinely Euclidean. But if we find the growth is slower or faster, the space possesses a positive or negative curvature.'

'So, are you saying that in the case of positive curvature the space has less volume within a given distance, and in the case of negative curvature more volume?' ventured Mr Tompkins.

'Just so,' smiled the professor.

'But that would mean, if space were positively curved—this space all around us here—then the volume of that beach ball is not $\frac{4}{3}\pi r^3$, but something smaller?'

'That's right. And if it's a case of negative curvature then it will be more. Mind you,' the professor added, ' with a sphere that small the difference would be minute; you'd never be able to detect it. Your only hope would be to measure over vast distances, like those one deals with in astronomy; that's why I was talking of the distances between galaxy clusters spread right throughout the Universe.'

'This is extraordinary,' murmured Mr Tompkins.

'Yes.' agreed the professor. 'But there's more to come. If the curvature is negative, we expect the three-dimensional space to extend indefinitely in all directions—like the two-dimensional saddle surface. On the other hand, a positive curvature would imply that three-dimensional space is finite and closed.'

'What does that mean?'

'What does that mean?' mused the professor. 'It would mean that if you took off vertically in a space rocket from the North Pole, and you continued in the same direction—in a straight line—eventually you would arrive back at the Earth, approaching it from the opposite direction, and landing at the South Pole.'

'But that's impossible!' exclaimed Mr Tompkins.

> 这里的讨论是有些让人头大，不过，记住二维的面和三维的空间的类比就行了，在三维的空间中如果曲率是正的，就像二维的球面，那么空间就是有限、封闭的。

'As impossible as an explorer circumnavigating the globe, always travelling exactly due West, assuming the Earth to be flat, so believing he is getting further and further away from his starting point—only to find himself back at the point where he started, approaching it from the East. And another thing...'

'Not *another*,' protested Mr Tompkins, his head already in a spin.

'The Universe is expanding,' continued the professor regardless. 'Those galaxy clusters I told you about are receding into the distance. The further off the cluster, the faster it is moving away. It's all due to the Big Bang. You've heard of the Big Bang, I take it?'

Mr Tompkins nodded, wondering where Maud had gone.

'Good,' his companion resumed. 'That's how the Universe began. There was a Big Bang with everything initially coming from a point. There was nothing before the Big Bang; no space, no time, absolutely nothing. That's when *everything* began. The clusters of galaxies are still flying apart in the aftermath of that gigantic explosion. But they *are* slowing down—due to the mutual gravitational forces between them. The crucial question is whether the clusters are moving apart fast enough to escape the pull of their gravity (in which case the Universe will expand for ever), or whether one day they will come to a halt, and thereafter get pulled back together. That would give rise to a Big Crunch.'

'What would happen then—after this Big Crunch?' asked Mr Tompkins, his interest once again aroused.

'Well, that might be *that*. The end. The Universe goes out of existence.Or it could rebound—a Big Bounce. It could be a Universe that is oscillating: expansion,followed by contraction, followed by a further cycle of expansion, and so on—for all time.'

'And what's it to be?' asked Mr Tompkins. 'Will the expansion go on for ever, or will it one day turn in to a

3 Mr Tompkins Takes a Holiday

Big Crunch?'

'Not sure. It depends on how much matter there is in the Universe—the matter producing the slowing-down gravitational force. It looks very finely balanced. The average density of matter is close to what we call the *critical value*—the limiting value that separates the two scenarios. It's hard to tell because we now know that most of the matter in the Universe is not luminous; it's not like the matter bound up in stars; it does not shine. We call it dark matter. Being dark, it's much harder to detect out there, but we know it makes up at least 99% of all matter-and it's that which brings the total density close to the critical value.'

> 对从事宇宙研究的科学家，事情就是那么巧，物质的密度偏偏就接近临界值。

'That's a shame,' commented Mr Tompkins. 'I would like to have known which way the Universe was going to go. What bad luck the density making it so difficult to decide.'

'Well...yes and no. The fact that the density has come out so close to critical (of all the possible values it could conceivably have taken on) raises the suspicion that there must be some deep underlying reason for it. Many people suspect that early on in the Big Bang there was some mechanism at work that automatically led to the density taking on that special value. In other words, it's no coincidence that the density comes out somewhere near the critical value; it doesn't just happen by chance; it actually *has to have* the critical value. In fact we think we know what that mechanism is. It's called *inflation theory*...'

> 从表面现象深入思考更深层的原因（用科学术语讲就是"机制" mechanism），这也是科学工作的典型特色。

'Jargon, Dad!'

The pair were startled by Maud's arrival. She had come up from behind them while they were still engrossed in their conversation. 'Give it a rest,'she said.

'In a minute,' the professor insisted. Turning back to his friend, he continued, 'I was just about to say—before we were so rudely interrupted—all these things we've been talking about are connected. If there is enough matter to cause a Big Crunch, then there will be enough

> 科学家习惯讲行话（jargon），但行话不适于普及，对这一点教授的女儿更有意识。

to produce positive curvature, and this will result in a closed Universe with a finite volume. On the other hand, if there is not enough matter...' He paused, gesturing to Mr Tompkins that it was his turn to take up the story.

'Er. If, as you say, there is not enough matter...er...' Mr Tampkins felt acutely embarrassed—not particularly about making a fool of himself before his teacher, but somehow the thought that Maud was listening intently made it much worse. 'Yes, as I was saying, if there's not enough matter to give you critical density, then the Universe will expand for ever, and...and...er, I'm just guessing...You'd get negative curvature...? and the Universe would be infinitely big...?'

'Excellent!' exclaimed the professor,'What a pupil!'

'Yes. Very good,' Maud agreed. 'But we all know the density is likely to be critical, so the expansion will eventually come to a halt—but only in the infinite future. I've heard it all before. Now are you coming for a dip?'

It was a while before Mr Tompkins realized that the question was addressed to him. 'Me?! You meant would I come for a swim?'

'Well. You don't think I meant *him*, do you?!' she laughed.

'Er, well I'm not dressed for it. I'll have to go and get my swimming trunks.'

'Of course. I had assumed you would be wearing *something*,' she said with a knowing look.

> 汤普金斯不想在教授女儿面前出丑,越急越说不清,结果却歪打正着。

> 这是教授女儿对汤普金斯愿意听他爸爸讲解的奖励吗?

4 The Notes of the Professor's Lecture on Curved Space

教授那篇关于弯曲
空间的演讲稿

Ladies and gentlemen:

Today's topic is curved space and its relation to the phenomena of gravitation.

There's clearly no problem imagining a curved line or a curved surface. But what could we possibly mean by a curved space—a curved *three-*dimensional space? It is obviously impossible to form a mental picture of what a curved three-dimensional space would look like. To do that one would have somehow to view it from 'outside' so to speak—from some other dimension (in the same way as we view the curvature of a two-dimensional surface by seeing how it extends into the third dimension). However, there is another approach to the investigation of curvature—a mathematical approach, rather than one relying on visualisation.

Take first of all curvature in a two-dimensional surface. We mathematicians call this surface curved if the properties of geometrical figures drawn on it are different from those on a plane. We determine the degree of the curvature by measuring the deviation from the classical rules of Euclid. For example, if you draw a triangle on a flat piece of paper the sum of its angles is equal to two right angles (as you know from elementary geometry). You can bend this paper to give it a cylindrical, a conical, or still more complicated shape, but the sum of the angles in the triangle drawn upon it will always remain equal to two right angles. The geometry of the surface, therefore,

> 又是一篇让人费神的学术演讲，读起来需要耐心，要思考，好在里面有不少直观的例子。

> curvature（曲率），这个重要的数学概念，是这篇演讲中核心的概念。

47

does not change with such deformations. From the point of view of the 'internal' or *intrinsic* curvature, the surfaces obtained are just as flat as a plane (even though we would commonly call them 'curved').

By way of contrast, you cannot fit the sheet of paper on to the surface of a sphere or a saddle—not without squashing or stretching it. This is because the geometry of the surface of a globe, say, is fundamentally different from that of a flat surface. Take for instance a triangle on a globe. To draw a triangle on this surface we would need the equivalent of three 'straight lines'. As on a flat surface, we define a 'straight line' on the curved surface to be the shortest distance between two points. That means we are dealing with arcs of great circles—great circles being the intersection of the spherical surface with planes drawn through the centre of the globe (for example, lines of longitude on the Earth are great circles). If you were to draw a triangle using such arcs, you would find the simple theorems of Euclidean geometry would no longer hold. In fact, a triangle formed, for example, by the northern halves of two meridians and the section of the equator between them, will have two right angles at its base and an arbitrary angle at the top—a sum clearly greater than two right angles.

On the other hand, with a triangle drawn on a saddle surface, you would find that the sum of its angles would always be *smaller* than two right angles.

Thus, to determine the curvature of a surface it is necessary to study the *geometry* on this surface. Merely looking at it from outside can be misleading. By looking you would probably place the surface of a cylinder in the same class as the surface of a globe. But as we have noted, the first is actually the same as a flat surface, and only the second is curved in the sense of having an *intrinsic* curvature. As soon as you get accustomed to this strict mathematical notion of curvature, you should have no difficulty in understanding what a physicist means in discussing whether the three-dimensional

4 The Notes of the Professor's Lecture on Curved Space

space in which we live is curved or not. It is unnecessary to get 'outside' the 3-D space to see whether it 'looks' curved. Rather, we remain within the space, and carry out experiments to see whether the common laws of Euclidean geometry hold or not.

注意，欧几里得几何学并不是唯一可能的几何学。

But you might be wondering why we should in any case expect the geometry of space to be anything other than 'commonsense' Euclidean. In order to show you that geometry can indeed depend on physical conditions, let us imagine a large round platform uniformly rotating around its axis like a turntable. Suppose small measuring rulers are placed end-to-end in a straight line along a radius from the centre to a point on the periphery. Additional rulers are placed around the periphery to form a circle.

common sense（常识）恰恰是在科学的发展中经常要被修正的东西。

According to an observer A, stationary relative to the room in which the platform is situated, the rulers placed around the periphery of the turntable are moving in the direction of their lengths as the turntable rotates. They will therefore be length-contracted (as we learned from the first lecture). It therefore takes *more* rulers to complete the circle than would have been necessary if the table had been stationary. The rulers lying along the radius, oriented so as to lie in a direction at right angles to the motion, will not undergo length contraction. It will therefore take the same number to span the distance from the centre of the table to its periphery regardless of the table's motion.

用这个旋转舞台的例子来说明问题，是伽莫夫的机智之处，在正规的课本中，是很难见到这样的示例的。

Thus, the distance measured round the circumference, C, (in terms of the number of rulers required) will be greater than the normal $2\pi r$, r being the measured radius.

在这里，几何规律已经不是通常的（normal），即不是欧几里得式的了。

As we have seen, all this makes perfect sense to observer A in terms of the length contraction produced by the motion of the rulers around the periphery. But what of an observer B, placed at the centre of the turntable and rotating with it? What will she make of it all? She would see the same number of rulers involved as did observer

It takes more rulers to complete the circle

A, and so would likewise conclude that the ratio of circumference to radius did not conform with Euclidean geometry. But suppose the platform were a closed room without windows, she would observe no motion. To what then would she attribute the unusual geometry?

Observer B might not know about the motion, but she would be aware that there was something odd about her surroundings. She would note that objects placed at different locations on the table do not remain stationary. They accelerate away from the centre, the acceleration being dependent on the distance of their location from the centre. In other words, they appear to be subject to a force(a centrifugal force). It is a peculiar force in that it causes all objects to accelerate from any particular location with identically the same acceleration regardless of mass. In other words, the 'force' appears to adjust its strength automatically to match the mass of the object, thus always producing the acceleration characteristic of the location. Observer B concludes that there must be some connection between this 'force' and the non-Euclidean geometry she finds.

> 这里清楚地表明了"力"与非欧几何之间的关系，或者说，在这种非欧几何空间中，会出现我们在"正常"情况下看不到的"力"。

4　The Notes of the Professor's Lecture on Curved Space

Not only that, consider the path taken by a light beam. For the stationary observer A, light always travels in straight lines. But suppose a beam were to skim across the surface of the rotating platform. Though it would continue to move in a straight line according to A, its path as traced out over the surface of the rotating platform would *not* be straight. This is because it takes a finite time for the light to cross the platform, and in that time, the table rotates through a certain angle. (It is as though you were to pull a sharp knife in a straight line across a rotating disc; the scratch on the surface would be curved rather than straight.) Thus, observer B at the centre of her rotating platform would find that a light beam passing from one side to the other would follow a curved, rather than straight, path. This phenomenon, like the one involving the circumference and the radius, she would have to attribute to the 'force' characterising the special physical conditions at work in her surroundings.

> 在这种空间中，光线也不再走直线。

This 'force' not only affects geometry, including the paths of light beams, but also the passage of time. This can be demonstrated by placing a clock on the periphery of the rotating platform. Observer B finds that it runs more slowly than a clock placed at the centre of her platform. This phenomenon is most readily understood from the point of view of the stationary observer A. As far as he is concerned the clock placed at the periphery is moving due to the table's rotation, and is thus time dilated compared with the clock at the centre, which remains at the same position. Observer B, not aware of the motion, must attribute the slowing down of the clock to the presence of the 'force'. Thus, we see that both geometry and the passage of time can be a function of physical circum-stances.

> 钟表也开始变得"不正常"。

> 此段的最后一句是一个重要的结论。

We now turn to a different physical situation—that which we find close to the surface of the Earth. All objects are pulled towards the centre of the Earth by the force of gravity. This can be regarded as somewhat similar to the way all objects placed on the rotating platform are pulled

51

towards the periphery. The similarity is strengthened when we further note that the acceleration undergone by the object is independent of its mass; it depends solely on the location. The correspondence between gravity and accelerated motion can be seen even more clearly in the following example:

Suppose a spacecraft floats freely somewhere in space so far from any stars that there is no force of gravity inside it. All objects inside such a craft, including the astronaut travelling in it, will thus have no weight and will float freely. Now the engines are switched on, and the craft gains velocity. What will happen inside? It is easy to see that, as long as the craft is accelerated, all the objects in its interior will show a tendency to move towards the back end of the craft—what we might call the 'floor'. To say the same thing in another way, the floor will be moving towards these objects. If, for example, our astro-naut holds an apple in her hand and then lets it go, the apple will continue to move (relative to the surrounding stars) with a constant velocity—the velocity with which the craft was moving at the moment when the apple was released. But the craft itself is accelerated; consequently the floor, moving all the time faster and faster, will overtake the apple and hit it. From this moment on, the apple will remain permanently in contact with the floor, being pressed to it by the steady acceleration.

For the astronaut inside, however, this will look as if the apple 'falls down' with a certain acceleration, and after hitting the floor remains pressed to it by its own 'weight'. Dropping different objects, she will notice furthermore that all of them fall with exactly equal accelerations (neglecting the friction of the air) and will remember that this is exactly the rule of free fall discovered by Galileo Galilei by dropping balls from the leaning tower of Pisa. *In fact the astronaut will not notice any difference between the phenomena in the accelerated cabin*

4 The Notes of the Professor's Lecture on Curved Space

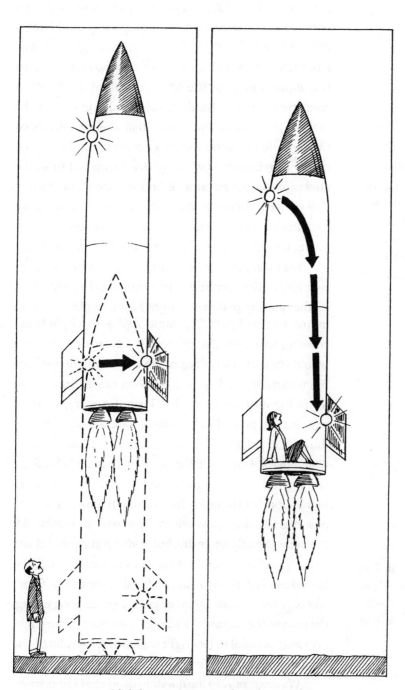

A light beam crossing an accelerated spacecraft

and the ordinary phenomena of gravity. If she chooses, she can use a clock with a pendulum, put books on a shelf without any danger of their floating away, and hang a picture on a nail in the wall. The picture might in fact be a portrait of Albert Einstein—the one who first indicated this equivalence of the acceleration of a system of reference on the one hand, and of a field of gravity on the other. It was on this simple basis that Einstein developed the so-called general theory of relativity. His special theory of relativity is what we dealt with last time; the effects on space and time of uniform constant motion. The general theory adds to this the effects on space and time of gravity. And, as I said, this is done through noting the equivalence of gravity and *accelerated* motion.

> 注意这里提到的"狭义相对论"（special theory of relativity）和"广义相对论"（general theory of relativity）的差别。

For example, take the case of a light beam. We noted that under the conditions of centrifugal acceleration on the rotating platform, a light beam would appear to follow a curved path. The same applies to a light beam crossing an accelerated spacecraft. An outside observer would see such a light beam move in a straight line. The beam starts off lined up with the point exactly opposite on the facing wall. If the craft had been stationary, it would have hit that point. But because of the craft's acceleration during the passage of the beam across the cabin, the far wall moves. As a result, the beam hits a point behind the one at which it had been originally aimed—a point closer to the 'floor' of the craft. The astronaut makes a similar observation: the beam originally starts out aiming for the point directly opposite, but ends up at a point closer to the 'floor' of the craft. As far as she is concerned, the beam followed a *curved* path and 'fell' towards the 'floor'. Not only that but she finds her geometry has gone wrong; the sum of the angles of a triangle formed by three light rays are not equal to two right angles, and the ratio of the circumference of a circle to its radius is not equal to 2π.

> 由此看到在这个例子中，也与前一个"旋转舞台"的例子一样，欧几里得几何不再成立。

We come now to the question of greatest importance. We have just seen that in an accelerated system of

4 The Notes of the Professor's Lecture on Curved Space

reference, not only do objects 'fall', but a light beam also 'falls' towards the 'floor', following a curved path. We therefore ask whether, in accordance with the equivalence principle, we are justified in concluding that light beams will be bent by *gravity*.

> 重要的限定条件是：加速参考系 (accelerated system of reference）.

In order to get a measure for the expected curvature of a light ray in the field of gravity, we consider how much bending we expect in the case of the accelerating spacecraft. If l is the distance across the cabin, then the time t taken by light to cross it is given by.

> 不用太担心，这段计算不很难。

$$t = \frac{l}{c} \tag{5}$$

During this time, the ship, moving with the acceleration g, will cover the distance L given by the following formula of elementary mechanics:

$$L = \frac{1}{2}gt^2 = \frac{1}{2}g\frac{l^2}{c^2} \tag{6}$$

Thus, the angle representing the change of the direction of the light ray is of the order of magnitude

$$\phi = \frac{L}{l} = \frac{1}{2}gl/c^2 \tag{7}$$

where angle ϕ is in radians (1 radian is about 57 degrees). We see that ϕ is greater the larger the distance l which the light has travelled in the gravitational field. Here, the acceleration g of the craft has, of course, to be interpreted as the acceleration due to gravity. If I send a beam of light across this lecture room, I can take l to be roughly 10 metres. The acceleration of gravity g on the surface of the Earth is 9.81m/s^2, and

$c = 3 \times 10^8$ m/s, so we get
$\phi = \frac{1}{2}(9.81 \times 10)/(3 \times 10^8)^2 = 5 \times 10^{-16}$ radians
$= 10^{-10}$ seconds of arc (8)

Thus you can see that the curvature of light can definitely not be observed under such conditions. However, near the surface of the Sun, g is 270 m/s^2, and the total path travelled in the gravitational field of the

Sun is very large. The exact calculations show that the value for the deviation of a light ray passing near the solar surface should be 1.75 seconds of arc. This is indeed the value observed by astronomers for the displacement of the apparent position of stars seen near the solar limb during a total eclipse, compared with their positions at night-time at other times of the year when the Sun is in a different part of the sky. Indeed, since the advent of astronomy using radio emissions from strongly emitting galaxies called *quasars*, one does not even need to wait for an eclipse; radio waves originating from quasars and passing close to the limb of the Sun can be detected without difficulty in broad daylight. It is these observations that give us our most precise measurements of the bending of light.

最初，正是英国科学家爱丁顿带领的观测队通过在日全食时进行的观测，证明了广义相对论的推论，为广义相对论的正确性提供了证明。

So we conclude that the bending of light that we found in the accelerated system does indeed apply equivalently to a gravitational field. What about the other strange effect our observer B found on the rotating platform—the one whereby a clock placed at some distance from her on the periphery of the platform was found to be running slow? Would this mean that a clock placed at some distance from us in a gravitational field would behave similarly? In other words, are the effects of acceleration and the effects of gravity not only very similar, but identical?

The answer to this can be given only by direct experiments. And, indeed, these do prove that time can be affected in an ordinary field of gravity. The effects predicted through the equivalence of accelerating motion and gravitational fields are very small: that is why they have been discovered only after scientists started looking specially for them.

也许正是因为效应很小，人们以往才认识不到。

Using the example of the rotating platform discussed before, we can easily estimate the order of magnitude of the expected change of the clock rate. It is known from elementary mechanics that the centrifugal force acting on a particle of unit mass, located at a distance r from the centre, is given by the formula

估算数量级（order of magnitude），这是物理学家常用的方法。

$$F=r\omega^2 \qquad (9)$$

where ω is the constant angular velocity of rotation of our platform. The total work done by this force while moving the particle out from the centre to the periphery is then

$$W=\tfrac{1}{2}R^2\omega^2 \qquad (10)$$

where R is the radius of the platform.

According to the above-stated equivalence principle, we have to identify F with the force of gravity on the platform. and W with the difference of gravitational potential between the centre and the periphery.

Now, we must remember that, as we have seen in the previous lecture, the slowing down of the clock moving with the velocity v is given by the factor

$$\sqrt{\left(1-\frac{v^2}{c^2}\right)}$$

This can be approximated by

$$1-\frac{1}{2}\frac{v^2}{c^2}+\ldots$$

If v is small as compared with c, we can neglect other terms. According to the definition of the angular velocity we have $v=R\omega$ and the 'slowing-down factor' becomes

$$1-\frac{1}{2}\left(\frac{R\omega}{c}\right)^2=1-\frac{W}{c^2} \qquad (11)$$

giving the change of rate of the clock in terms of the difference of gravitational potentials at the places of their location.

So, if we imagine placing one clock on the ground and another on the top of the Eiffel tower (about 300 metres high) the difference of potential between them will be so small that the clock on the ground will go slower only by a factor 0. 999,999 ,999,999,97 compared with that at the top.

> 300米高的埃菲尔铁塔带来的影响，实在是小得可以忽略不计。

In fact, an experiment carried out by R. V. Pound and G. A. Rebka has demonstrated this small effect by examining the difference in the rates of atomic vibrations at the top and bottom of a tower 22.5 metres high. The same effect has also been found by comparing the rates of atomic clocks flown in aircraft with those on the ground. Agreement with observation is obtained only if, in addition to the time dilation caused by the aircraft's motion(special relativity), one takes account of the slowing down of the Earth-bound clock compared with the high flying one due to the difference in gravitational potential.

Considerably larger effects than these, however, are to be found once one involves the much stronger gravity of the Sun. The difference of gravitational potential between the surface of the Earth and the surface of the Sun is much larger, giving a slowing down factor of 0. 999, 999,5. This is much easier to measure, and provided the first confirmation of these ideas. Of course, nobody can place an ordinary clock on the surface of the Sun and watch it go! The physicists have much better means. By using a spectroscope we can observe the periods of vibration of different atoms on the surface of the Sun and compare them with the periods of the atoms of the same elements put into the flame of a Bunsen-burner in the laboratory. The vibrations of atoms on the surface of the Sun should be slowed down by the factor given by equation(11), and as a consequence, the light emitted or absorbed by them should have a somewhat lower frequency than in the case of terrestrial sources, i.e. the frequencies should be shifted towards the red end of the spectrum. This *gravitational red-shift* has been observed in the spectra of the Sun, and of several other stars, and the results agree with the value given by our theoretical formula. This shows that the processes on the Sun really do take place somewhat more slowly than they do on Earth, owing to the difference in gravitational potential.

These observations have therefore demonstrated the equivalence of the effects of acceleration and those of gravitation. So, with this in mind, let me now return once

4 The Notes of the Professor's Lecture on Curved Space

more to the curvature of space:

You remember that we came to the conclusion that the geometry obtained in accelerating systems of reference is different from that of Euclid, and that such spaces should be considered as curved spaces. Since any gravitational field is equivalent to some acceleration of the system of reference, this means also that any space in which the gravitational field is present is a curved space. Or, going a step further, we can say that *a gravitational field is just a physical manifestation of the curvature of space.*

We know that gravity arises in the vicinity of masses. Thus, we would expect that the curvature of space at each point should be determined by the distribution of masses, and would reach maximum values close to heavy objects. I cannot enter into the rather complicated mathematical system describing the properties of curved space and their dependence on the distribution of masses. I will mention only that this curvature is in general deter-mined not by one, but by ten different numbers which are usually known as the components of gravitational potential, $g\mu v$, and represent a generalization of the gravitational potential of classical physics which I have previously equated with W in equation (10). Corres-pondingly, the curvature at each point is described by ten different radii of curvature usually denoted by $R\mu v$. Those radii of curvature are connected with the distribution of masses by the fundamental equation of Einstein:

$$R_{\mu v} - \frac{1}{2} g_{\mu v} R = -8\pi G T_{\mu v} \qquad (12)$$

where R is another kind of curvature, and the source term $T_{\mu v}$ (representing the *cause* of the curvature) depends on densities, velocities and other properties of the gravitational field produced by masses. G is the familiar gravitational constant.

对于普通读者，知道有这个方程就可以了，知道空间的曲率与引力有关就可以了,其他可暂不深究。

59

> 其实，对广义相对论的实验直接验证并不很多，对水星的解释是其中很重要的一个。

This equation has been tested out, for example, by studying the motion of the planet Mercury—the planet closest to the Sun, and hence the one with the orbit most sensitively dependent on the details of Einstein's equation. It is found that the perihelion of the orbit (i. e. the point of closest approach to the Sun of the planet as it executes its elongated elliptical path) does not remain fixed in space, but is found with each turn of the orbit to have systematically shifted its orientation relative to the Sun. Part of this precession is attributable to the perturbing gravitational fields of the other planets, and part can be explained in terms of the special relativistic increase in mass due to the planet's motion. But there remains a tiny residual amount of 43 seconds of arc per century which cannot be accounted for by the old Newtonian theory of gravity, but finds a ready explanation in terms of general relativity.

This observation, together with the other experimental results I have mentioned in this lecture, confirm us in our judgement that general relativity is the theory of gravity that best fits what we actually see happening in the Universe.

Before ending this lecture, allow me to indicate two further interesting consequences of equation(12):

If we consider a space uniformly filled with masses, as, for example, our space is filled with stats, galaxies and clusters of galaxies, we must conclude that, apart from localized large curvatures near particular stars or galaxies, space should possess an *overall* curvature due to the combined effect of all the masses—a regular ten-dency to curve uniformly over large distances. Mathe-matically there are different solutions. Some of them correspond to space finally closing in upon itself, and thus possessing a finite volume, somewhat similar to a sphere. The others represent a curved space, but not curved sufficiently to cause closure; instead, the space is infinite in extent, having no boundaries—rather analogous to the saddle surface I mentioned at the beginning of this lecture.

A second important consequence of equation (12) is that such curved spaces should be in a state of steady expansion or contraction. This physically means that the particles (the galaxy clusters) filling the space should be flying away from each other, or, on the contrary, approaching each other. Further, it can be shown that for a closed space with a finite volume, the expansion phase will be followed by a contraction phase (with possibly further expansion and contraction phases to follow—thus giving rise to an oscillating Universe). On the other hand, an infinite expanding 'saddle-like' space would continue to expand for ever.

The question of which of these different mathematical possibilities corresponds to the space in which we live is very much a live issue at present. It can only be resolved by experimental observation on the movements of the galaxy clusters (including their rate of slowing down);either that or by accounting for all the mass present in the Universe and calculating how great the slowing down effect will be. At present, the astronomical evidence is unclear. Though it is certain that we are at present in an expanding phase, whether this will ever turn into a contraction (and consequently, whether the space is finite or infinite in size) is not yet definitely settled.

> 其实像这种未知的答案，在科学中太常见了，不过，这里提到的问题，却关系着未来——很久很久以后的未来——人类的命运。

5 Mr Tompkins Visits a Closed Universe

汤普金斯先生访问一个封闭宇宙

> 虽然部分地是由于教授女儿的吸引力，让他学习物理，这么多的信息对于汤普金斯先生可实在是有些负担过重，让他的大脑疲倦并乱成一团了。

That evening in the Beach Hotel, the professor and his daughter were deep in conversation. They talked freely of both cosmology and art. Mr Tompkins joined in from time to time as best he could, but for the most part was happy just to observe and listen.
He was fascinated by Maud; he had never met anyone like her. But in due course he became sleepy and made his excuses. Climbing the stairs and reaching his room, he quickly changed into his pyjamas and collapsed on to the bed, pulling the blanket over his head. His tired brain was all mixed up.

As he lay there, one thought kept recurring. The type of cosmology that really intrigued him was that of a closed Universe—the one where if you go off from the North Pole in a straight line you will end up at the South Pole. At least it would be a Universe with a finite volume (he simply could not get his mind round the infinite volume of an open Universe). Fair enough, the professor seemed to have his reasons for thinking that the density of matter had the critical value, and so you would not be able to make that odd type of journey, and the expansion would not give way to a contraction and a Big Crunch. But what if he were wrong? What if there was a lot more dark matter out there than they had yet accounted for? What if…?

> 关于暗物质（dark matter），最近又成为宇宙学研究的热门话题。

These thoughts were interrupted as he became aware that he was uncomfortable. He had the strange feeling that instead of lying on a comfy spring mattress he was

5 Mr Tompkins Visits a Closed Universe

HE BECAME AWARE THAT HE WAS UNCOMFORTABLE

stretched out on something hard. He pulled back the blankets and peeped out. To his astonishment he found himself lying on a slab of rock out in the open. The hotel had vanished!

> 又开始做梦了。

The rock was covered with some green moss, and in a few places little bushes were growing from cracks in the stone. The space above him was illuminated by some glimmering light and was very dusty. In fact, there was more dust in the air than he had ever seen, even in the films representing dust storms in the American midwest. He tied his handkerchief round his nose to keep from breathing the dust.

But there were more dangerous things than the dust in the surrounding space. Occasionally stones, the size of his head and larger, came whirling through space, hitting the ground around him. He also noticed one or two rocks, about 10 metres across he judged, floating through space at some distance away.

Another strange thing was that there appeared to be no distant horizon—despite his being perched high up. He decided he had better explore his surroundings. So it was he began crawling over the surface. Because the rock curved down quite sharply, he held on grimly to the protruding edges in constant fear of falling off. But then he gradually became aware of something odd; although he had moved down onto a very steep part of the rocky face—so steep he could now no longer see the blanket he had left behind, he did not feel any tendency to fall; he was still being pulled securely onto the surface. Emboldened, he continued crawling. Eventually he reckoned he must have gone through about $180°$—in other words he ought to be directly *underneath* his starting point—and still there was no tendency to fall off into the surrounding dusty depths of space. He was presumably now upside-down compared with when he started out. It was then it dawned on him that the rock he was on had no visible means of support. It was a planet! A tiny planet similar

> 当一块大石头成为一个小行星时，当然就不存在支撑的问题了。

5 Mr Tompkins Visits a Closed Universe

to the floating rocks he had seen.

To his great surprise and relief it was at that moment he almost bumped into the legs of a familiar figure. It was the professor. He was standing there busily noting down observations in a note-book.

'Oh, it's you,' observed the professor casually. 'What are you doing down there? Lost something?'

Mr Tompkins sheepishly let go of his hand-hold, and gingerly stood up. To his great relief, not only did he not fall off into space, he did not even feel as though he would drift off into space. He began to understand what was going on. He remembered that he was taught in his schooldays that the Earth is a big round rock moving freely in space around the Sun. Everything is pulled towards its centre, so there is no danger of 'falling off,' no matter where you are positioned on its surface. Now he was gently but firmly being pulled towards the centre of this new 'planet'—a planet so tiny its population numbered two.

'Good evening,' said Mr Tompkins; 'What a relief to see you.'

The professor raised his eyes from his note-book. 'There are no evenings here, ' he said. 'There is no Sun,' and with that he returned again to his note-book.

夜晚概念的存在是以太阳为前提的。

Mr Tompkins felt uneasy; to meet the only living person in the whole Universe, and to find him so preoccupied! Unexpectedly, one of the little meteorites came to his help. With a crashing sound, the stone hit the book in the hands of the professor and knocked it hard. It flew up into space away from their little planet. 'Oh dear,' said Mr Tompkins, 'I hope that wasn't important. I don't reckon our gravity is strong enough to pull it back.' As they watched, the book continued its journey into the furthest depths of space, getting smaller and smaller.

'Not to worry,' replied the professor. 'You see, the space in which we are now is not infinite in its extension. Oh I know that you were doubtless taught in school that space is infinite, and that two parallel lines never meet. This, however, is not true for the space of this particular Universe—the one we are now in. Our normal Universe is, of course, very large indeed; about 100,000,000,000,000,000,000,000 kilometers across at present, which for most purposes is fairly infinite. If I had lost my book there, it would have taken an incredibly long time to come back—even assuming it were a Universe of the closed type that this one is. Here, however, the situation is rather different. Just before the note-book was torn out of my hands, I had figured out that this space is only about five miles in diameter, though it is expanding. I expect the book back in not more than half an hour.'

'Are you saying that the book is going to do one of those round trip journeys in a straight line,' ventured Mr Tompkins. 'Like the one you said about taking off from the North Pole…'

'…and landing back at the South Pole? Yes,' replied the professor. 'Precisely. The same thing is going to happen to my book—unless it's hit on its way by some other stone and gets deflected from the straight track.'

'And this has nothing to do with the gravity of our little planet pulling it back?'

'No, nothing at all to do with that. As far as the gravity here is concerned, the book has escaped into space. Here, take these binoculars, and see if you can still see it.'

Mr Tompkins put the binoculars to his eyes, and through the dust which somewhat obscured the whole picture, he managed to see the professor's note-book traveling through space far, far away. He was somewhat surprised by the pink colouring of all the objects, including the book, at that distance. Not only that ,'Your book is returning already,' he cried out excitedly. 'Yes, yes, it's definitely growing larger now.'

'No, no,' said the professor, 'it'll still be going away. Here, give those to me.' He took back the binoculars, and looked intently. 'No, as I said, it's still going away. The fact

5 Mr Tompkins Visits a Closed Universe

that it appears to be growing in size—*as if* it were coming back—that's due to a peculiar focusing effect on the rays of light due to the closed, spherical nature of the space.'

He lowered the binoculars and scratched his graying head.'How can I put it…? Yes. Suppose we were back on Earth, and let's imagine that horizontal rays of light (aimed at the horizon) could be kept going all the time hugging the curved surface of the Earth (say, by refraction of the atmosphere). Under those circumstances, if an athlete were to run away from us, it wouldn't matter how far she went, we would be able, using powerful binoculars, to see her all the time during her journey. Now, if you think about the globe, you will see that the straightest lines on its surface, the meridians, first diverge from one pole, but, after passing the equator, begin to converge towards the opposite pole. If the rays of light traveled along the meridians, you, located for example at one pole, would see the person going away from you growing smaller and smaller only until she crossed the equator. After this point you would see her growing larger; it would seem to you that she was returning, albeit going backwards. Once she reached the opposite pole, you would see her as large as if she were standing right by your side. You would not be able to touch her, of course, just as you cannot touch the image formed by a spherical mirror.

'Right now, the professor continued, 'this behaviour of light as it travels over the two-dimensional curved surface of the Earth can be used as an analogy for how light rays behave in this strangely curved three-dimensional space we find ourselves in. In fact I do believe the image of the book is about to arrive.'

> 对于理解高维空间中的事，用低维空间做类比总是相对简单有效的办法。

As he said that, the image of the book appeared to be only a few yards away, and coming closer. It was big enough now for one no longer to need the binoculars to see it by. However, it looked rather odd; the contours were not sharp, but seemed washed out, and the writing on the cover could hardly be recognized; the whole book looked like a photograph taken out of focus and underdeveloped.

'You can see now it's only an image—not the real thing,' said the professor. ' See how its badly distorted by the light having had to travel halfway across the Universe. And notice how you can see other little planets behind the book—through its pages.'

Mr Tompkins reached out and tried to grab the 'book' as it sped passed, but his hand simply passed through the image without encountering any resistance.

'No, no,' admonished the professor. 'The book itself is now very close to the *opposite* pole of the Universe. As I've said, what you see here is just an image—in fact two images of it. The second image is just behind you and when both images momentarily coincided just then, that was when the real book was exactly at the opposite pole.'

Mr Tompkins didn't hear; he was too deeplyabsorbed in his thoughts, trying to remember how the images of objects are formed in elementary optics by concave and convex mirrors and lenses. When he finally gave up, the two images were receding in opposite directions.

'And all these strange effects are due to the matter in the Universe?' he eventually asked.

'That's right. The matter we're standing on—our tiny planet—curves the space in our immediate vicinity, and it is this that is responsible for the way we are held onto its surface. But more than that, the gravity of this planet combines with that of all the other masses in the Universe to produce the overall curvature that gives rise to these lensing effects. In fact, in general relativity one dispenses altogether with talk of gravitational 'forces' as such,and simply thinks in terms of curvature.'

'But tell me, if there were no matter, would we have the kind of geometry I was taught at school, and would parallel lines never meet?'

'That's right,' answered the professor, 'but neither

> 这么一会就过了半个宇宙，这个宇宙可真够小的。

5 Mr Tompkins Visits a Closed Universe

would there be any material creature to check it.'

In the meantime, the image of the book went off again far away in the original direction, and started coming back for the second time. Now it was still more damaged than before, and could hardly be recognized at all, which, according to the professor, was due to the fact that this time the light rays had travelled round the whole Universe.

'And if we pop round to the other side of our planet...' he added, grabbing Mr Tompkins by the arm and marching him the few yards it took to get to the other side. 'There,' he declared, pointing in the opposite direction. 'There. Can you see? Here comes my book. It's about to complete its journey round the Universe.' With a triumphant grin, he stretched out his hand, caught the book, and pushed it into his pocket. 'The trouble with this Universe is that there is so much dust and stones around, it makes it almost impossible to see round the world. Notice these shapeless shadows around us? Most probably they're the images of ourselves, and surrounding objects. It's just that they're so distorted by dust and irregularities of the curvature of space that I cannot even tell which is which.'

> 在这样的世界里，凡尔纳要写幻想小说，恐怕就不是关于环游地球，而是环游宇宙了。

'Does the same effect occur in our normal Universe- the one we used to live in?' asked Mr Tompkins.

'Probably not—not if we're right about the density being critical. But,' the professor added with a twinkle in his eye.

'you have to admit, it's still fun to think this kind of thing through, don't you agree?'

By now the sky had considerably changed. There seemed to be less dust about, So Mr Tompkins was able to take off the handkerchief from around his face. The small stones were passing much less frequently and hitting the surface of their planet with much less energy. Not only that, but the other planets had drifted much farther away by now and could hardly be seen at this distance.

'Well, I must say life is getting a lot less scary,' he

commented, 'Though I must say it's become quite chilly.' He picked up the blanket and wrapped it round him. 'Can you explain the change in our surroundings?' he asked, turning to the professor.

'Very easily; our little Universe is expanding and since we have been here its radius has increased from five to about a hundred miles. As soon as I found myself here, I noticed this expansion from the reddening of the distant objects.'

对"红移"的形象想像!

'Ah. I did notice everything was pink at great distances,' said Mr Tompkins, 'but why does that signify expansion?'

'Oh that's not difficult to see.' said the professor. 'I take it you've noticed that the siren of an approaching ambulance sounds very high, but after the ambulance passes you, the tone is considerably lower? This is the so-called Doppler Effect: the dependence of the pitch (or frequency of the sound) on the velocity of the source. When the whole of space is expanding, every object located in it moves away with a velocity proportional to its distance from the observer. Therefore, the light emitted by such objects is of lower frequency, which in optics corresponds to redder light. The more distant the object is, the faster it moves and the redder it seems to us. In our normal Universe, which is also expanding, this red-dening, or the *cosmological red-shift* as we call it, permits astronomers to estimate the distances of the very remote galaxies. For example, one of the nearest galaxies. the Andromeda galaxy, shows a 0.05% reddening; this corresponds to the distance which can be covered by light in eight hundred thousand years. But there are also galaxies just on the limit of present telescopic power which show a reddening of about 500%, corresponding to distances of approximately ten thousand million light years (a 'light year' being—as the name implies—the distance travelled by light in one year). Such light was emitted when the Universe was less than a fifth its present size.

5 Mr Tompkins Visits a Closed Universe

The present rate of expansion is about 0. 000, 000, 0 1% per year. Our little Universe here grows comparatively much faster, gaining in size by about 1% per minute.'

'Will the expansion of this Universe here ever stop?' asked Mr Tompkins.

'Of course it will ,' said the professor. 'I told you in one of the lectures that a closed Universe like this one would entail the expansion eventually coming to a halt, this then being followed by the contraction phase. For a Universe this small the expansion phase should, I reckon, last no more than a couple of hours.'

'A couple of hours,' echoed Mr Tompkins. 'But that would mean there can't be long to go before…' His voice trailed off as the implication sank in.

> 这里的梦境几乎就像是关于宇宙理论的"理想实验"——一种在思维中的逻辑演练。

'Yes,' murmured the professor. 'I think we are now observing the state of largest expansion. That's why it's become so cold.'

In fact, the thermal radiation filling up the Universe, and now distributed over a very large volume, was giving only very little heat to their planet; the temperature was at about freezing-point.

'It's lucky for us that there was originally enough radiation to give some heat even at this stage of expansion,' the professor added. 'Otherwise it might become so cold that the air around our rock would condense into liquid and we would freeze to death.'

> 在这里又悄悄地引入了宇宙背景辐射的概念。

He peered intently through his binoculars once more. 'Ah, yes,' he said after a while. 'The contraction has already begun. It'll soon be warm again.'

He offered the binoculars to Mr Tompkins, who took them and scanned the heavens. He noticed that all the distant objects had changed their colour from pink to blue. This, according to the professor, was due to the fact that all the stellar bodies had started moving towards them. He also remembered the analogy given by the professor of the high pitch of the whistle of an approaching train.

> 宇宙的表现恰恰是冷胀热缩。

Rubbing himself to get warm, he commented, 'Well, I'll be glad when it heats up again.' But then a thought struck him. He turned anxiously to the professor. 'If everything is contracting now, shouldn't we expect that soon all the big rocks filling the Universe will come together and that we shall be crushed between them?'

'I wondered how long it would take you to work that out,' answered the professor calmly. 'But not to worry. Just think: well before that happens, the temperature will rise so high that we shall be vapourised! I suggest you just lie down and observe as long as you can.'

'Oh my!' moaned Mr Tompkins. 'I am beginning to feel hot already, even in my pyjamas.'

> 原来科幻也可以成为汤普金斯先生的噩梦。

It was not long before the hot air became unbearable. The dust, which became very dense now, was accumulating around him, and he felt as if he were being choked. He struggled to free himself from the blanket, when suddenly his head emerged into cool air. He swallowed a deep breath.

'What's happening?' he called out to the professor—only to discover that his companion was no longer with him. Instead, in the dim light of morning, he recognised the hotel bedroom. Sighing with relief he disengaged himself from his blanket; it had become entangled after what must have been a very restless night.

> 还是我们这个现实的宇宙好!

'Thank God we're still expanding!' he muttered, as he made his way to the bathroom. 'That's what you might call a close shave,' he thought as he reached for the razor.

6 Cosmic Opera
宇宙之歌

It was the final evening of their holiday, and Mr Tompkins and Maud were taking one last stroll along the beach by the water's edge. Was it really only a week since they had first met? Though at first he had been quite nervous of speaking with her, he being shy by nature, they knew each other well enough now for the conversation to flow easily. He found it extraordinary that one person should have such wide interests. Not only that, he was delighted to note that she seemed to take as much pleasure being with him as he with her. He could not possibly think why. Except that on one occasion the professor had let slip that his daughter had been badly let down in the past; her engagement to some high-flying executive had been abruptly broken off. Perhaps she just felt safe with him and his rather humdrum, but reassuringly secure life.

He looked up at the Milky Way. 'I must say your father has opened up a whole new world for me. It's sad how most people seem to go through life without ever appreciating just how extraordinary the world is.'

Picking up a handful of pebbles, he lazily aimed them at a rock sticking up out of the water. Then he shot a quick glance at her. 'Why won't you show me your sketches?'

'I've told you. They're not the sort you show anyone. They're *working* sketches—ideas. Just ideas. That's all. They try to capture the *feel* of the place. They wouldn't mean anything to you. It's only when I get back to the studio and work on them something emerges–or not, as the case maybe.'

教授确实是向汤普金斯先生，也向广大的读者展示了一个全新的世界。

'Then, can I come and visit your studio one day when we get back?' he asked.

'Of course,' she replied. 'I'd be disappointed if you didn't.'

By now they had got back to the hotel. Mr Tompkins ordered drinks, and for the last time they sat on the patio looking out to sea.

'Your father told me there was a time when you were cut out for a career in physics,' he commented.

'Oh, I wouldn't say that,' she laughed. 'Wishful thinking. That was what *he* wanted.'

'Yes, but you were good at physics, weren't you?' he persisted.

She shrugged. 'Yes. You could say that.'

'So why...?'

'Why?' she repeated wistfully. 'Oh, I don't know. Rebellious teenager, I suppose. That and the fact that it wasn't easy in those days for a girl to show an interest in science. Biology maybe, but not physics. Peer pressure and all that. It's different now—well, at least it's not quite so bad now.'

'But how come you still know so much physics after all this time?'

'Oh I don't really. Forgot most of it long ago. Except for astronomy and cosmology. Now *that* I have tried to keep up with. Which reminds me...' she looked at him in amus-ement.

'Reminds you of what?' he asked.

'Fancy taking me to the opera?'

'Opera!' he exclaimed. 'What...what do you mean? What's opera got to do with anything?'

'Oh it's not a *real* one,' she added with a laugh. 'No. it's an amateur one. It was written ages ago by someone who used to be in Dad's department. It's all about the Big Bang theory versus the Steady State theory...'

'Steady State? What's that?' he enquired.

'The Steady State theory says that the Universe did

> 可惜作者没有恰当的性别视角，否则，就此应该可以演绎出更多，更有深度寓意的故事，现在，已经有人专门在研究性别和物理学的关系了。

6 Cosmic Opera

not begin with a Big Bang...'

'But we know it *did*. Your father's told me all about the expansion of the Universe—the way all the galaxies are still flying apart in the aftermath of the big explosion,' Mr Tompkins protested.

'Ah, but that doesn't prove anything. You see there were these physicists, Fred Hoyle, Hermann Bondi and Tommy Gold, who suggested that the Universe could keep on renewing itself. As fast as galaxies moved away, new matter was created in the spaces left behind. This collected together to form new stars and galaxies, which in their turn moved apart, making room for yet more matter, and so on.'

'So, how did all this get started,' asked Mr Tompkins, clearly intrigued.

'Oh, it didn't. There was no start, no beginning. It has always been going on, and always will. It's a world with no beginning and no end. That's why it was called the Steady State theory; the world looks essentially the same at all times.'

'Hey, I like the sound of that,' enthused Mr Tompkins. 'Yes, it's got the right...the right kind of *feel* about it. You know what I mean? Somehow the Big Bang idea doesn't have that appeal. You find yourself asking why it was supposed to have happened at that particular instant in time; why not some other instant? It seems so... so *arbitrary* somehow. Now if there's *no* beginning...'

'Hold on! Hold on!' interrupted Maud, 'Don't get too carried away. The Steady State theory is dead. Dead as a dodo.'

'Oh,' said Mr Tompkins disappointedly. 'Why's that? How can they be so sure?'

But before Maud could reply, her father emerged from the hotel doorway to remind her that they had to make an early start home in the morning. As she took her leave of Mr Tompkins, he hurriedly asked, 'But what about the opera?'

在类似这样的问题上，宇宙学和形而上学的界限就很难分清了。

> 即使是一个如今已不居主流地位的科学理论，在科学史的意义上，因其曾作为科学发展过程进展的一部分，也是值得纪念的。比如，就像这次演出为了纪念稳态宇宙理论50周年。

'Oh yes,' she said. 'Saturday evening, 8 o'clock, in the main physics lecture theatre—the one you normally go to for Dad's lectures. The department is reviving the *Cosmic Opera*. It's just a bit of fun. The 50th anniversary of the first proposal of the Steady State theory; I think that's the excuse. See you there.' With that she followed her father into the hotel, briefly turning her head to blow Mr Tompkins a playful goodnight kiss.

there was a good turnout for the performance. The theatre was almost full as Mr Tompkins took his place, accompanied by the professor and Maud.

'You'd better take a look at your programme, 'Maud remarked to him.'Quickly. Before they put the lights out. You won't understand who the characters are if you don't.'

He rapidly scanned the typed sheet he had been handed at the door. He just managed to get to the end of the background notes when the theatre was plunged into darkness, and the six-piece orchestra, squashed together in a small space to the side of the raised platform, launched into the prelude, *precipite volissimevolmente*. To the accompaniment of raucous applause from the students making up most of the audience, the makeshift curtains erected around the platform were drawn back. Everybody immediately had to shield their eyes, So brilliant was the illumination of the stage. The intensity was such as to fill the entire theatre in one brilliant ocean of light.

> 强光象征着宇宙"大爆炸"的开端。

'That idiot technician! He'll blow every fuse in the building!' muttered the professor fiercely under his breath. But it was not to be. Gradually the 'Big Bang' brilliance faded, leaving a darkened space, illuminated by a multitude of rapidly rotating Catherine wheels. Presumably these were supposed to be the galaxies that formed some time after the Big Bang.

'Now they're set on burning the place down,' the professor fumed. 'I should never have given them permission for this nonsense.'

6　Cosmic Opera

Maud leant across and tapped his arm, drawing attention to the fact that the 'idiot technician' was in fact standing discreetly at the corner of the stage, holding a fire extinguisher at the ready in case it was needed. Meanwhile the students were loudly oooohing and ahhhing like small children at a firework festival, before being shushed up at the entry of a man in a black cassock and a clerical collar. According to the program me, this was AbbéGeorges Lemaitre from Belgium, the first man to propose the Big Bang theory of the expanding Universe.Singing with a thick accent, he began his aria:

在此歌剧中：人物、服装，都以象征的方式对应着不同的理论观点及其提出者。

读者也不妨照乐谱试着唱唱这些咏叹调，也挺好听。

O, Atome prreemorrdiale!

All-containeeng Atome!

Deessolved eento fragments exceedeengly small.

Galaxies forrmeeng,

Each wiz prrimal enerrgy!

O, rradioactif Atome!

O, all-containeeng Atome !

O, Univairrsale Atome—

Worrk of z'Lorrd!

77

Z'long evolution
Tells of mighty firreworrks
Zat ended een ashes and smouldairreeng weesps.
We stand on z'ceendairres
Fadeeng suns confrronteeng us,
Attempteeng to rremembairre
Z'splendeur of z'origine.
O, Univairrsale Atome—
Worrk of Z'lorrd!

> 这时登场的，实际上就是本书作者伽莫夫，当然，按照他对宇宙学的贡献，他是完全有资格在此成为主角一展歌喉的。

Father Lemaitre having finished his aria to riotous cheers from the student element of the audience (which had clearly spent the earlier part of the evening in the bar), there now appeared a tall fellow who (according to the programme again) was the physicist George Gamow, born in Russia, but later settled in the USA. He took centre stage and began to sing:

Finished his aria to riotous cheers

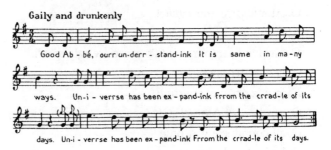

Gaily and drunkenly

Good Ab-bé, ourr un-derr-stand-ink It is same in ma-ny ways. Un-i-verrse has been ex-pand-ink Frrom the crrad-le of its days. Un-i-verrse has been ex-pand-ink Frrom the crrad-le of its days.

Good Abbé, ourr underrstandink
It is same in many ways.
Univerrse has been expandink
Frrom the crradle of its days.
Univerrse has been expandink
Frrom the crradle of its days.

You have told it gains in motion.
I rregrret to disagrree,
And we differr in ourr notion
As to how it came to be.
And we differr in ourr notion
As to how it came to be.

It was neutrron fluid—neverr
Prrimal Atom, as you told.
It is infinite, as everr
It was infinite of old.
It is infinite, as everr
It was infinite of old.

On a limitless pavilion
In collapse, gas met its fate,
Yearrs ago (some thousand million)
Having come to densest state.
Yearrs ago (some thousand million)
Having come to densest state.

All the Space was then rresplendent
At that crrucial point in time.
Light to matterr was trranscendent
Much as meterr is, to rrhyme.
Light to matterr was trranscendent
Much as meterr is, to rrhyme.

Forr each ton of rradiation
Then of matterr was an ounce,

> 伽莫夫很擅长写打油诗,在他的唱词中,也流露出打油诗的风格。

Till the impulse t'warrd inflation
In that grreat prrimeval bounce.
Till the impulse t'warrd inflation
In that grreat prrimeval bounce.

Light by then was slowly palink.
Hundrred million yearrs go by...
Matterr, over light prrevailink,
Is in plentiful supply.
Matterr, over light prrevailink,
Is in plentiful supply.

Matterr then began condensink
(Such are Jeans' hypotheses).
Giant, gaseous clouds dispensink
Known as prrotogalaxies.
Giant, gaseous clouds dispensink
Known as prrotogalaxies.

Prrotogalaxies were shattered,
Flying outward thrrough the night.
Starrs werre forrmed frrom them, and scatterred
and the Space was filled with light.
Starrs werre forrmed frrom them, and scatterred
And the Space was filled with light.

Galaxies arre everr spinnink,
Starrs will burn to final sparrk,
Till ourr univerrse is thinnink
And is lifeless, cold and darrk.
Till ourr univerrse is thinnink
And is lifeless, cold and darrk.

> 由稳态宇宙理论的提出者霍依耳登场，用咏叹调来介绍其理论，真的很通俗。

It was then the turn of Fred Hoyle. He suddenly materialised from nowhere in the space between the brightly shining galaxies. He pulled a Catherine wheel out of his pocket, and lit it. As it began to rotate, he triumphantly held the newly born galaxy aloft and launched into his aria:

The Universe, by Heaven's decree,
Was never formed in time gone by,
But is, has been, shall ever be—
For so say Bondi, Gold and I.
Stay, O Cosmos, O Cosmos, stay the same!
We the Steady State proclaim!

The aging galaxies disperse,
Burn out, and exit from the scene.
But all the while, the Universe
Is, was, shall ever be, has been.
Stay, O Cosmos, O Cosmos, stay the same!
We the Steady State proclaim!

And still new galaxies condense
From nothing, as they did before.
(Lemaitre and Gamow, no offence!)
All was, will be for evermore.
Stay, O Cosmos, O Cosmos, stay the same!
We the Steady State proclaim!

But even as the Hoyle character sang, one could not help but notice that despite his inspiring hymn to the unchanging nature of the Cosmos, by now most of the little 'galaxies' had fizzled out.

And so the opera continued to its final act when the entire cast assembled to sing the final rousing chorus:

'Your years of toil,'
Said Ryle to Hoyle,
'Are wasted years, believe me.
The steady state
Is out of date.
Unless my eyes deceive me,

My telescope
Has dashed your hop;
Your tenets are refuted.
Let me be terse:
Our Universe
Grows daily more diluted!'

Said Hoyle, 'you quote
Lemaitre, I note,
And Gamow. Well, forget them!
That errant gang
And their Big Bang—
Why aid them and abet them?

You see, my friend,
It has no end
And there was no beginning,

> 在想像中（也许是因为太难或太复杂，这段合唱没有给出乐谱），这种歌剧中多声部的合唱倒真适合用来表现不同观点的"争鸣"。

6 Cosmic Opera

As Bondi, Gold,
And l will hold
Until our hair is thinning!'

'Not so!' cried Ryle
With rising bile
And straining at the tether;
'Far galaxies
Are, as one sees,
More tightly packed together!'

'You make me boil!'
Exploded Hoyle,
His statement rearranging;
'New matter's born
Each night and morn.
The picture is unchanging!'

'Come off it, Hoyle!
I aim to foil
You yet' (The fun commences)
'And in a while,'
Continued Ryle,
'I'll bring you to your senses!'

At the conclusion there was thunderous applause, stamping, and a standing ovation to rival the headiest night at Covent Garden. Eventually the temporary curtain got stuck in the closed position, so pre-empting any further curtain calls. The audience then dispersed—the younger element making their way back to the student union bar.

'Doing anything special tomorrow, Maud?' asked Mr Tompkins as he was about to take his leave.

'Not really, she replied. ' You could come round to my place for coffee if you like. 11 o'clock suit you?'

> 作者伽莫夫用歌剧的形式表现各种关于宇宙的理论，确有独到之处，不过，这也许与他年轻时在哥本哈根求学的经历有关，那里的物理学家当时就有用戏剧来表现（或者说影射）不同科学家的传统。

7 Black Holes, Heat Death, and Blow Torch

黑洞、热寂和喷灯

'I suppose this must be it,' muttered Mr Tompkins consulting the scribbled map Maud had given him. There was no name on the gate to confirm that this was indeed Norton Farm. At the end of the drive he could see a huge, rambling farm house. It as not what he  was expecting, but he decided he'd better go and ask. It was then he spotted Maud. She was crouching down, weeding a flower bed. They greeted each other warmly.

'Quite some place you've got here,' he said admir-ingly. 'I didn't know painting paid so well. Aren't you supposed to starve in a garret somewhere, and suffer for your art?'

At first she looked puzzled, but then suddenly burst out laughing.'All this mine?!' she exclaimed. 'Some hopes! No, it's all been split up—ever since the Nortons left. It's separate units now. This is my bit.' She indicated an extension that had been more recently added. 'Come on in and make yourself at home.'

While they waited for the kettle to boil she took him on a quick tour of her small, but very pleasantly furnished home. Then settling down on the sofa in the living room, they had their coffee and biscuits.

> 这倒有些英式的在"下午茶"时间讨论学术的味道。

'So. What did you make of the opera yesterday?' she asked.

'Oh, it was great fun,' he said. 'I didn't get all the allusions of course. But no, I enjoyed it. Thanks for suggesting it. The only thing was…'

'Yes?'

'Well, it was just that when I got home I couldn't

7 Black Holes, Heat Death, and Blow Torch

help wondering what happened to the Steady State theory. It seems such a *sensible* kind of theory.'

'Don't let Dad hear you say that,' she smiled. 'It took a lot of persuading to get him to allow that opera to take place. He didn't want the students to get confused. He makes a big song and dance about science being based on *experiment – not* aesthetics. It doesn't matter how attracted you are to a theory, if the experimental results go against it, then you ditch it.'

'And is the evidence against it as strong as you seemed to be making out the other day?' he asked.

'Oh yes,' she replied. 'The evidence is overwhelmingly in favour of the Big Bang. In the first place we know that the Universe has changed over time—we can *see* that it has changed.'

Mr Tompkins frowned. 'See it?'

'Yes. You have to remember that the speed of light is finite; it takes time for light to reach us from a distant object. When you look far out into space you also look far back into time. That light from the Sun, for example,' she said, looking out of the window, 'that took eight minutes to get here, so that means we are seeing the Sun as it was eight minutes ago—not as it is at this precise moment. The same goes for more distant objects, such as the galaxy in the constellation of Andromeda. You must have seen photographs of that; it's in all books on astronomy. That galaxy is located about one million light-years away, so the photos show how it looked one million years ago.'

'So what's your point?'

'The point is this:' she continued, 'Martin Ryle observed that the number of galaxies in a given volume went up the further he probed into space–in other words, the further he looked back in time. And that of course is what you would expect if the Universe were gradually thinning out with time; it would have been denser in the past.'

'They mentioned that towards the end of the opera, didn't they?' Mr Tompkins asked.

> 虽然科学家也经常强调科学理论的美学特征，但审美的要求毕竟是第二位的，以经验（实验、观察）为基础才是第一位的。

> 这里所谈的，就是对大爆炸理论给予支持的观察证据之一。

'That's right. Not only that, but we now know that the nature of galaxies themselves has changed over time. There was a phase, soon after they formed after the Big Bang, when they burned much brighter than they do today; they're called *quasars* when they behave like that. Quasars are seen only at great distances—meaning they existed long ago and not now—which again does not fit in with the idea of an unchanging Universe.'

'OK, I'm beginning to be convinced,' he admitted.

'But I'm not finished,' she persisted. 'Take the primordial nuclear abundances.'

'The *what*?'

'The ratio of the different kinds of particle coming out of the Big Bang. You see, at an early stage of the Big Bang, everything was hot; everything was moving about fast and smashing into each other. At that stage all you had were sub-nuclear particles (neutrons and protons), electrons, and other fundamental particles. You didn't have the nuclei of heavy atoms; no sooner did one form(through neutrons and protons fusing together), than it got smashed up again—in a subsequent collision. It was only later, as the Universe expanded and cooled down, that the collisions became less violent; only then

> 这里作者是借慕德之口，把有关的证据和逻辑推论过程相对通俗地一一展示出来。

It's in all books on astronomy

7 Black Holes, Heat Death, and Blow Torch

could the newly formed nuclei survive. So that's how you got this *primordial nucleosynthesis*—as it's called.

'But this was not something that could go on indefinitely—nuclei absorbing more and more protons and neutrons, building up bigger and bigger nuclei,' she continued. It was a race against time. The temperature was still going down. That meant eventually it would be so low the nuclear particles would no longer have had enough energy to fuse. Not only that, but the *density* was going down too—because of the expansion—so the collisions were getting less and less frequent. So, for a combination of these reasons, the point was reached where the nuclear reactions came to a halt and the mix of heavy nuclei no longer changed. That mix is called the *freeze-out mix*. And that mix of nuclei determines the mix of the different types of atom that were eventually to form.

'Now, the interesting thing is this,' she concluded. 'If you know what the density of matter in the Universe is today, you can work out what it must have been at any earlier stage, and in particular what it must have been at the time of primordial nucleosynthesis. And that in turn means you can work out theoretically what the freeze-out mix should have been. It turns out that you would expect 77% of the mass to be in the form of hydrogen (the lightest element), 23% helium (the next lightest), and just traces of heavier nuclei. And that is exactly what is observed today when you examine the atomic abundances of the interstellar gases!'

> 注意这里的一个词："从理论上"（theoretically）！科学就是这样在观察和理论之间互动着。

'OK, you win,' Mr Tompkins conceded. 'The Big Bang wins the day.'

'But I haven't told you the most convincing evidence of all,' added Maud, growing more enthusiastic by the moment.

'You're beginning to sound like your father.'

Ignoring the remark, she continued. 'The cosmic

background radiation. You see, if the Big Bang was hot it would have been accompanied by a fireball—in the same way as a nuclear bomb going off gives out a blinding flash of light. Now the question is what has happened to all that radiation from the Big Bang. It must be in the Universe somewhere; there's no other place for it to be. All right, it won't be a blinding light anymore; it will have cooled down by now. By this stage it should have wave lengths in the microwave region. In fact, Gamow(remember him from last night?), well he calculated that it should have a wave length spectrum corresponding to a temperature somewhere in the region of 7 K, that is 7 degrees above absolute zero. And he was right; the remnants of the fireball have now been found. The radiation was discovered in 1965, purely accidentally, by two communications scientists, Penzias and Wilson. It has a temperature of 2. 73 K, which of course is very close to that predicted by Gamow.'

> 这也许是对大爆炸理论最重要的观察支持，尤其是，在一开始，它的发现并不是专门为寻找支持大爆炸理论的证据而进行的，但这种意外的发现，却加重其支持的力量，可以视为是对理论的"预言"的验证，这种对理论的预言的验证，是理论家最希望看到的。

Mr Tompkins said nothing; he was lost in thought. Maud looked at him quizzically.

'Was that all right…?' she asked. 'Convinced?'

Mr Tompkins shook himself out of his reverie. 'Oh yes. Yes, fine. That was fine. Thanks.But…'

'But what?'

'Well, it's just that I've got this mental picture of hydrogen and helium, and electrons, and radiation coming out of the Big Bang—and nothing else. So how come the world is the way it is today? Where did the Sun and the Earth come from? And what about you and me;we're not made out of just hydrogen and helium?'

'You're asking for 12,000 million years of history! How long have I got?'

'Would three minutes be enough?' asked Mr Tompkins hopefully.

She laughed. 'I'll give it a try. Ready?'

'Wait a moment.' he said,looking a this watch. 'OK. Away you go.'

7 Black Holes, Heat Death, and Blow Torch

'Right. A few minutes after the Big Bang we have hydrogen and helium nuclei and electrons. After 300,000 years things have cooled down enough that the electrons can now attach themselves to the nuclei. That gives us the first atoms. So now we have space filled with a gas. The gas is pretty uniform in density, but there are some slight in homogeneities—places where the density is somewhat greater or less than the average. The gas now starts to clump together around the dense spots—due to their extra gravity. The more gas they collect, the stronger their gravity becomes, sand the better they get at pulling in yet more gas from the surroundings. This gives us clouds of gas separating out from each other. Now inside each cloud little eddies or whirlpools form. These squash down and heat up. (This always happens when you squash a gas down into a smaller volume; it heats up.) Eventually the temperature gets so high it ignites nuclear fusion processes—and that's how stars are born. So, after about 1 billion years we've got our galaxies of stars. (Actually, it might have happened the other way round. Instead of the galaxy cloud forming first and then breaking down into stars, the stars might have formed first and then later collected together to form a galaxy. No-one knows for certain at present.) One way or the other, we have our stars. These are powered by the nuclear fusion processes. These not only release energy, but also build up the nuclei of the heavier atoms—the atoms we shall later be needing to make the Earth, and the stuff of our bodies. Eventually the stellar nuclear fires run out of fuel. For a medium-sized star like the sun that takes about 10,000 million years. It swells up in its old age to become what is called a *red giant* star. Then it shrinks down to a *white dwarf,* and this then slowly fizzles out to become a cold cinder. More massive stars end their active lives in a much more spectacular fashion. They go out with a bang—literally. A *supernova explosion*. It's this explosion that throws out some of the

> 以下的一大段描述，就像是按大爆炸理论编写的宇宙演化科普影片的脚本。

newly synthesized nuclear material—the heavy nuclei. These are now mixed up with the in-terstellar gas and can now collect together to form a second generation star, and for the first time, rocky planets like the Earth (which of course weren't there for the first generation of stars). It's then that evolution by natural selection takes over, converting the chemicals on the surface of the planet into you and me. That's how we come to be made of stardust!'

Maud stopped abruptly. 'There! That's it! How long did that take?'

Mr Tompkins grinned. 'Well, that was just over two minutes…'

'Good,' announced Maud. 'That gives me a minute to talk about black holes.'

'*Black holes*?'

'Yes. That's what's left after one of these really massive stars blows up. It flings out some of the material, as I said, but the rest collapses down to form a black hole.'

'What exactly is a black hole?' asked Mr Tompkins. 'I've heard of them, of course…'

'A black hole is what you get when the gravity force is so strong nothing can resist it. All the matter of the star collapses down to a point.'

'A *point*!' exclaimed Mr Tompkins. 'Do you really mean that? An actual *point*?'

'That's right. No volume,' was the reply. 'The point where all the matter is concentrated is surrounded by a region of incredibly strong gravitational field. It's so strong, if you get within a certain distance of it (within its *event horizon*), nothing can get away from it—not even light. That's why it's black. Anything that finds itself within the event horizon gets sucked into that point at the centre.'

'Amazing!' murmured Mr Tompkins. 'And what lies beyond the black hole?'

'Beyond? Who knows? There doesn't have to be anything 'beyond'. The stuff that's fallen in just stays

7 Black Holes, Heat Death, and Blow Torch

there at the centre. Oh, there are various speculations that it might all travel through a tunnel of some sort, linking our Universe to some other. It then comes squirting out into the other Universe as a 'white hole'. But that's all it is: pure speculation.'

'And are we sure that black holes actually do exist?'

'Oh yes. The evidence is pretty strong. Not only black holes from collapsed old stars, but at the centre of galaxies too—massive ones that might have swallowed up a mill-ion or so stars.'

Mr Tompkins beamed at Maud admiringly.

'Why are you looking at me like that?' she asked curiously.

'Oh nothing. I was just wondering how on Earth you know all this?'

she shrugged modestly. 'Dun no. Mostly from those, I suppose. ' She nodded at a shelf full of popular science magazines.

> 为什么在地球上能知道这些？读科普书，这只是慕德这样的非专业人士给出的表面的答案，更深刻的问题也许是：为什么人类竟能有这样的认识和理论思考能力？这个问题就不那么好回答了。

'One last question Einstein,' he asked. 'How is it all going to end? What's going to happen to the Universe in the future? Your Dad said something about it expanding forever, but slowing down to a halt in the infinite future.' 'That's correct—if the the inflation theory is right, and the density of matter in the Universe has the critical value. By then all the nuclear fuel will have been used up, the stars will all have died, many will get sucked into the black hole at the centre of their galaxy, the Universe will become freezing cold and lifeless. *The Heat Death of the Universe*—that's what they call it.'

Mr Tompkins shuddered. 'Not sure I like the sound of that.'

'Oh I don't know. I shouldn't let it bother you,' she responded brightly. 'We'll all be dead and buried long before that. Anyway, that's enough of that. Let's change the subject.'

> 还好，科学文献中通常将 Heat Death 译为"热寂"，看上去不像把 Death 直译为"死"那么可怕。

'Yes. I'm sorry. What must you think of me?'

'Oh that's all right. Make the most of it while you

can,' she chuckled. 'I won't be any use to you next week.'

'Next week? What's happening next week?'

'Dad's getting on to quantum theory in his next lecture, isn't he?'

'I believe so.'

'Well, I never did fathom quantum theory. All I can say is: Best of luck! But right now it's my art work. Did you seriously want to see some of it?'

'Your work? Of course I did,' he replied. 'Where do you keep it? Is it far to your studio?'

'Far? No. It's just across the courtyard. I've got the use of an old barn out there. That was why I came to Norton Farm in the first place. It wasn't the house, it was the barn I wanted.'

Maud's studio was a wonderland. Mr Tompkins had never seen anything like it. Her creations (you couldn't call them paintings) were extraordinary. Although they were framed and were meant to be hung on a wall, they were made out of all sorts of things: plaster, wood, metal tubing, slate, pebbles, tin cans, etc. The various items were stuck together in elaborate and powerful collages.

> 慕德的艺术创作是 Creations，但是，人们形容上帝创世，或宇宙创生时，也用 Creation 这个词。

'They're wonderful,' he exclaimed. 'I had no idea. Quite wonderful. Mind you,' he continued hesitantly, 'I can't say I *understand* them—not really understand them. But I do *like* them,' he added emphatically.

She smiled. 'They aren't physics theories, you know. They're not there to be *understood*. You have to *experience* them.'

> 科学与艺术的重要差别：艺术家更关心体验，或者说感受（experience）。

For a while he stood silently contemplating one of the works. Then he ventured, 'You have to develop a two-way relationship with it—an interaction. It's not complete until the viewer puts something of himself into it—sees it in relation to some experience of his own. Is that what you mean?'

She shrugged noncommittally. 'That's my latest,' she said, nodding at the one he was examining. 'What do you see in that one?'

7 Black Holes, Heat Death, and Blow Torch

'This one? A beach. Things washed up on a beach. Gnarled and time-worn; each with its own private history to tell; now brought together, by chance, at the same place, at the same time.'

She was regarding him closely. It was a look he hadn't noticed before, and he immediately felt rather foolish.

'Sorry. I'm talking nonsense. Probably been reading too many exhibition catalogues. One of the advantages of working up in town is that I get the chance to visit galleries and art exhibitions in my lunch break, 'he explained. 'I like art—some of it, at any rate. I try to keep abreast of what's going on.'

She smiled.

'Tell me,' he continued 'How do you get this weathered effect? It almost looks as though it's been salvaged from a fire.' He pointed to some charred-looking pieces of wood embedded in the plaster.

慕德的 Creation 是用喷灯来完成的。那么，上帝的 Creation 又是用什么"工具"完成的呢？也许，这种以与人类相类比的提问方式就不对。

She gave him a mischievious look. 'I'll show you, if you like—but you'll have to watch out for yourself.'

With that she struck a match and set light to a blow torch standing on a nearby table. Picking it up she played the roaring flames over the face of one of the pictures. It was not long before the wooden parts of it were glowing red. In no time the studio was filled with smoke. Backing away in some alarm, Mr Tompkins found the door and threw it open to allow the smoke to escape. From there he watched in fascination. He caught sight of Maud's face. It was a picture of total concentration. It was at that moment he realized that he was in love.

8 Quantum Snooker

量子台球

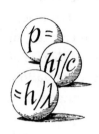

The audience for the professor's next lecture was not quite as large as it had been at the beginning of the series;some had clearly not stayed the course.But it was still sizeable. As Mr Tompkins sat there waiting, he recalled Maud's remark about the difficulties of learning quantum theory; he anxiously won-dered how he would manage.But he was determined to master it, if he could.He even held out hopes that where quantum physics was concerned, it might prove to be *his* turn to tutor *her*!

The professor entered…

Ladies and gentlemen:
In my two previous lectures I tried to show you how the discovery of the upper limit for all physical velocities brought us to a complete reconstruction of nineteenth century ideas about space and time.

This development of the critical analysis of the foundations of physics did not, however, stop at this stage; still more striking discoveries and conclusions were in store. I am referring to the branch of physics known as *quantum theory*. This is not so much concerned with the properties of space and time themselves, as with the mutual interactions and motions of material objects taking place in space and time.

In classical physics it was always accepted as self-evident that the interaction between any two physical bodies could be made as small as required by the conditions of the experiment, and practically reduced to

教授的这次演讲，几乎就是一部简明的量子理论发展史，而量子理论却是与人们通常更熟悉的经典物理学理论截然不同的。

zero whenever necessary. For example, suppose the aim is to investigate the heat developed in a certain process, and one is concerned that the introduction of a thermometer would take away a certain amount of the heat, and thus introduce a disturbance. The experimenter was confident that by using a smaller thermometer (or a very tiny thermocouple), this disturbance could be reduced to a point below the limits of needed accuracy.

The conviction that any physical process can, in principle, be observed with any required degree of precision, without disturbing it by the observation, was so strong that nobody troubled to formulate such a proposition explicitly. However, new empirical facts accumulated since the beginning of the twentieth century steadily brought physicists to the conclusion that the situation is really much more complicated. In fact, *there exists in nature a certain lower limit of interaction which can never be reduced*. This natural limit of accuracy is negligibly small for most kinds of processes we are normally familiar with in ordinary life. They do, however, become highly significant when it comes to the interactions taking place in such tiny mechanical systems as atoms and molecules.

In the year 1900, the German physicist Max Planck was thinking about the conditions of equilibrium between matter and radiation. He came to a surprising conclusion: No such equilibrium was possible if the interaction between the matter and radiation took place continuously, as had always been supposed. Instead, he proposed that the energy was transferred between matter and radiation *in a sequence of separate 'shocks'*. A particular amount of energy was transferred in each of these elementary acts of interaction. In order to get the desired equilibrium, and to achieve agreement with the experimental facts, it was necessary to introduce a simple mathematical relation which stated that the amount of energy transferred in each shock was proportional to the frequency of the radiation responsible for the transfer of energy.

Thus, denoting the coefficient of proportionality by the symbol '*h*'. Planck was led to accept that the minimal portion, or *quantum*, of energy transferred was given by the expression

$$E = hf \qquad (13)$$

where *f* stands for the frequency of the radiation. The constant h has the numerical value 6.6×10^{-34} Joule second, and is usually called *Planck's constant*. (I take it you are all familiar with the notation that 10^{-34} means: 1/10,000,000,000,000,000,000,000,000,000,000,000 where that is supposed to be 34 zeros on the denominator.) Note that Planck's constant has a very small value, and it is this that is responsible for the fact that quantum phenomena are usually not observed in our everyday life.

> 这个被命名为"普朗克常数"的 h，是物理学中最重要的物理常数之一。

> 按日常的标准看，这个量实在是太小了。

A further development of Planck's ideas was due to Einstein. He concluded that not only is radiation emitted as 'packages of energy', but these packages subsequently transfer energy to matter in the same localized way as particles do. In other words, each package remains intact— it does not disperse its energy over a wide region, as was previously assumed. These packages of energy are referred to as 'quanta of light', or *photons*.

> 爱因斯坦正是为此工作而获得了诺贝尔奖——尽管人们现在认为这还不是他最重要的科学贡献。

In so far as photons are moving, they should possess, apart from their energy *hf*, a certain momentum also. According to relativistic mechanics, this should be equal to their energy divided by the velocity of light, *c*. Remembering that the frequency of light is related to its wave length, λ, by the relation $f = c/\lambda$, we can write for the momentum of a photon:

$$p = hf/c = h/\lambda \qquad (14)$$

Thus, a photon's momentum decreases with wave length.

One of the best experimental proofs of the correctness of the idea of light quanta, and the energy and momentum ascribed to them, was given by the investigation of the American physicist Arthur

Compton. Studying the interactions between light and electrons, he arrived at the result that electrons set in motion by the action of a ray of light behaved exactly as if they had been struck by a particle with the energy and momentum given by equations (13) and (14). The photons themselves, after colliding with the electrons, were also shown to suffer certain changes (in their frequency), in excellent agreement with the prediction of the theory.

Thus, we can say that, as far as the interaction with matter is concerned, the quantum property of electromagnetic radiation, such as light, is a well established experimental fact.

A further development of the quantum ideas was due to the Danish physicist Niels Bohr. In 1913, he was first to express the idea that *the internal motion of any mechanical system may possess only a discrete set of possible energy values*. As a result, that internal motion can change its state only by finite steps, the transition being accompanied by the radiation, or absorption, of a discrete amount of energy (the energy difference between the two allowed energy states). This idea was prompted by the observation that when atomic electrons emit radiation, the resulting spectrum is not continuous but consists of certain frequencies only—a 'line spectrum'. In other words, in accordance with equation (13), the emitted radiation has only certain energy values. This could be understood if Bohr's hypothesis concerning the allowed energy states of the emitter were correct—in this case, the energy states of the electrons in the atom.

The mathematical rules defining the possible states of mechanical systems are more complicated than in the case of radiation and we will not enter here into their formulation. Suffice it to say that, when describing the motion of a particle such as an electron, there are circumstances where it becomes necessary to ascribe to it the properties of a *wave*. The necessity for doing this was first indicated by the French physicist Louis de Broglie, on

> 玻尔，是一位普通人不太熟悉但却非常值得关注的物理学家。

the basis of his theoretical studies of the structure of the atom. He recognized that whenever a wave finds itself in a confined space, such as a sound wave within an organ pipe, or the vibrations of a violin string, only certain frequencies or wave lengths are permitted to it. These waves have to 'fit' the dimensions of the confining space, giving rise to what we call 'standing waves'. De Broglie argued that if the electrons in an atom had a wave associated with them, then because the waves were confined (to the vicinity of the atomic nucleus), their wave length would only be able to take on the discrete values permitted to standing waves. Furthermore, if this proposed wave length were related to the electron's momentum by an equation similar to equation (14) for light, i.e.

$$p_{\text{particle}} = h/\lambda \qquad (15)$$

then that would entail the momentum (and hence the energy) of the electron being able to take on only certain permitted values. This would of course provide a neat explanation of the discrete energy levels of electrons in atoms, and the consequent line spectrum nature of their emitted radiation.

In the following years the wave properties of the motion of material particles were firmly established by numerous experiments. They showed such phenomena as the *diffraction* of a beam of electrons passing through a small opening, and *interference phenomena* taking place even for such comparatively large and complex particles as molecules. The observed wave properties of material particles were, of course, absolutely incomprehensible from the point of view of classical conceptions of motion. De Broglie himself was forced to a rather unnatural point of view that the particles are 'accompanied' by certain waves which, so to speak, 'direct' their motion.

Due to the extremely small value of the constant, *h*, the wave lengths of material particles are exceedingly small—even for the lightest fundamental particle, the

> 按照德布罗意的观点，一切物质都有"波动"的性质，这又是一个与过去的"常识"相悖的观点。

electron. Whenever the wave length of radiation is small compared with the dimensions of apertures it might be going through, the diffraction effects are tiny and the radiation to all intents and purposes passes through in an undeviated manner. That is why a football passing through the gap between the goal posts does not undergo any visible change of direction due to diffraction. The wave nature of particles is of importance only for motions taking place in such small regions as the inside of atoms and molecules. Here they play a crucial part in our knowledge of the internal structure of matter.

One of the most direct proofs of the existence of the sequence of discrete states of these tiny mechanical systems was given by the experiments of James Franck and Gustav Hertz. They bombarded atoms with electrons of varying energy, and noticed that definite changes in the state of the atom took place only when the energy of the bombarding electrons reached certain discrete values. If the energy of the electrons was brought below a certain limit, no effect whatsoever was observed in the atoms. This was because the amount of energy carried by each electron was not sufficient to raise the atom from the first allowed quantum state into the second.

So how are we to view these new ideas in relation to classical mechanics?

> 当我们反思这些从日常观察经验得出的概念时，发现其实它们是"有问题"的，问题就来自下面要分析的"量子效应"。

The fundamental concept concerning motion in classical theory is that a particle, at any given moment, occupies a certain position in space and possesses a definite velocity characterizing its positional changes with time along its trajectory. These fundamental notions of position, velocity and trajectory, on which the entire edifice of classical mechanics is built, are formed (as are all our other notions) from observation of the phenomena around us. As we saw earlier with the classical notions of space and time, we must expect that they might be in need of radical modification as soon as our experience extends into new and previously unexplored regions.

8 Quantum Snooker

If I ask you why you believe that a moving particle occupies at any given moment a certain position, and that over the course of time it will describe a definite line called the trajectory, you will probably answer: 'Because I see it this way, when I observe its motion.' But let us analyse this method of forming the classical notion of the trajectory and see if it really will lead to a definite result. For this purpose we imagine a physicist supplied with sensitive apparatus, trying to pursue the motion of a little material body thrown from the wall of her laboratory. She decides to make her observation by 'seeing' how the body moves. Of course, to see the moving body, she must illuminate it. Knowing that light in general produces a pressure on the body, and so might disturb its motion, she decides to use short flash illumination only at the moments when she makes the observation. For her first trial she wants to observe only ten points on the trajectory and thus she chooses her flashlight source so weak that the integral effect of light pressure during ten successive illum-inations should be within the accuracy she needs. Thus, flashing her light ten times during the fall of the body, she obtains, with the desired accuracy, ten points on the trajectory.

注意,"看"这个日常动作背后的东西。

Now she wants to improve on the experiment and get a more precise fix on the trajectory—one hundred points this time. She knows that a hundred successive illuminations will disturb the motion too much and therefore, preparing for the second set of observations, chooses her flashlight ten times less intense. For the third set of observations, desiring to have one thousand points, she makes the flashlight a Hun dred times fainter than originally. Proceeding in this way, and constantly decreasing the intensity of her illumination, she can obtain as many points on the trajectory as she wants, without increasing the possible error above the limit she had chosen at the beginning. This highly idealized, but in principle quite possible, procedure represents the strictly logical way to construct the motion of a trajectory

这是一种构思巧妙的"理想实验"。

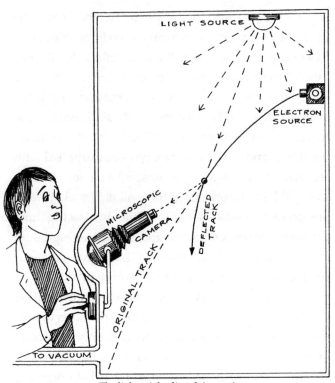

The light might disturb its motion

by 'looking at the moving body'. As you can see, from the point of view of classical physics, all of this is perfectly possible.

But now let us see what happens if we introduce the quantum limitations and take into account the fact that the action of any radiation can be transferred only in the form of photons. We have seen that our observer was constantly reducing the amount of light illuminating the moving body. But now we have to recognize that she will find it impossible to continue to do this once she has come down to a level of illumination equivalent to one photon per flash. Either all or none of the photon will be reflected from the moving body; one cannot have a fraction of a photon.

这几乎是在想像中的"看"了。

Now, the experimenter might argue that, in accordance with equation (14), the effect of a collision with a photon would be less if the wave length were larger. Consequently she resolves to increase the number of observations by making compensatory increases in the wave length of the light used.But here she meets with another difficulty. It is well known that when using light of a certain wave length, one cannot see details smaller than the wave length used. (One cannot paint a Persian miniature using a house-painter's brush!) Thus, by using longer and longer waves, she will spoil the estimate of each single point and soon will come to the stage where each estimate will be uncertain by an amount comparable to the size of her laboratory and more. Thus she will be forced finally to a compromise between the large number of observed points and the uncertainty of each estimate.Thus, she will never be able to arrive at an exact trajectory as a mathematical line such as that obtained by her classical colleagues. Her best result will be a rather broad, washed-out band.

The method discussed here is an optical method; we could try another possibility, using a mechanical method.For this purpose our experimenter can devise some tiny mechanical recording devices, say little bells on springs, which would register the passage of material bodies if such a body passes close to them. She can spread a large number of such bells through the space through which the moving body is expected to pass and after the particle's passage, the 'ringing of bells' will indicate its track. In classical physics one can make the bells as small and sensitive as one likes, and in the limiting case of an infinite number of infinitely small bells, the notion of a trajectory can be again formed with any desired precision. However, the quantum limitations for mechanical systems will spoil the situation again. The clappers of the bells are in the confined space of the bell itself. They will therefore have only certain discrete energy states allowed to them. If the bells are too small, the amount of momentum they need to take from the moving body in order to get the clapper to ring will be large, and as a result, the motion of

> 虽然这里换了机械的方法，是"听"声音，道理却与"看"光（量子）的反射是一样的，都属于理想中的"观察"。

the particle will undergo a correspondingly large disturbance. If on the other hand, the bells are large, little disturbance will be caused, but the uncertainty of each position will be large. The final trajectory deduced will again be a spread-out band!

These considerations might lead you to seek yet some other practical method for determining the trajectory—perhaps a more eladorate and complicated one. But let me point out that what we have been discussing here has not been so much the analysis of two particular experimental techniques, but an idealization of the most *general* question of physical measurement. Any scheme of measurement whatsoever will necessarily be reducible to the elements described in these two methods, and will finally yield the same result: exact position and trajectory have no place in a world subject to quantum laws…

> 有些枯燥抽象的演讲终于让汤普金斯先生又睡着了。

It was at this point in the lecture, Mr Tompkins gave up his battle to keep his leaden eyes open. His head drooped, suddenly to jerk up as he tried to force himself to keep awake; it drooped once more, another slight jerk,

Little bells on springs

8 Quantum Snooker

another droop…

Having ordered a pint from the bar, Mr Tompkins was about to find himself a seat, when his attention was taken by the click of snooker balls. He remembered that there was a snooker table in the back room of this pub, so he thought he would go and take a look. The room was filled with men in shirt sleeves, drinking and chatting animatedly, as they waited their turn to play. Mr Tompkins approached the table and started to watch the game.

There was something very strange about it! A player put a ball on the table and hit it with the cue. Watching the rolling ball, Mr Tompkins noticed to his great surprise that the ball began to 'spread out'. This was the only expression he could find for the odd behaviour of the ball which, as it moved, seemed to become more and more washed out, losing its sharp contours. It looked as if not one ball was rolling across the table but a great number of balls, all partially penetrating into each other. Mr Tompkins had often observed analogous phenomena before, but not on the strength of less than one drink. He could not understand why it was happening now.

'Hmm,' he thought, 'I wonder how this "fuzzy" ball is going to hit another one.'

The player who had hit the ball was evidently an expert; the moving ball hit the other head-on just as it was meant to; there was a loud sound of impact just like the collision between two ordinary balls. Then both the ball that had been moving and the one that had been stationary (Mr Tompkins could not positively say which was which) sped off 'in all different directions at once'. Very peculiar. There were no longer two balls looking only somewhat fuzzy, but instead it seemed that innumerable balls, all of them very vague and fuzzy, were rushing about within an angle of 180°round the direction of the original impact. It resembled a wave spreading from the point of collision, with a maximum

像台球这样的例子，在古典物理学中也常用，可是用到说明量子现象时，就让人匪夷所思了。

其实，严格地讲，这种可见的"弥散"景象只是对"概率波"的一种形象比喻，我们实际上是不可能"看"到概率波的，但这里的比喻真的是很形象。

flow of balls in the direction of the original impact.

'That's a nice example of probability waves they've got there,' said a familiar voice behind him. Mr Tompkins swung round to find the professor at his shoulder.

'Oh,it's you,' he said. 'Good. Perhaps you could explain what's going on here.'

'Certainly. The landlord seems to have got himself some balls suffering from "quantum-elephantism"—if I may so express myself. All objects in nature, of course, are subject to quantum laws. But Planck's constant (the quantity governing the scale of the quantum effects) is very, very small—at least it is normally.But for these balls here, the constant seems much larger—about ONE, I reckon.Which is actually quite useful; here you can see everything happening with your very own eyes. Normally you can only infer this sort of behaviour using very sensitive and sophisticated methods of observation.'

The professor became thoughtful. 'I must say I would dearly love to know how the landlord got hold of these balls.Strictly speaking, they can't exist in our world.Planck's constant is the same for all objects.'

'Maybe he imported them from some other world,' suggested Mr Tompkins. 'But tell me,why do the balls spread out like this?'

'Oh. that's all to do with the fact that their position on the table is not quite definite. You cannot indicate the position of a ball exactly; the best you can say is that the ball is "mostly here" but "Fartially somewhere else".'

'It actually, *physically* is in all these different places at once?' asked Mr Tompkins incredulously.

The professor hesitated, 'Maybe, maybe not. That's certainly how some people would say it was. Others would say that it is our *knowledge* of the ball's position that's uncertain. The interpretation of quantum physics has always been a subject for debate. There's no consensus even now.'

Mr Tompkins continued to gaze in wonder at the

fuzzy snooker balls. 'This is all very unusual,' he murmured.

'On the contrary,' insisted the professor, 'it is absolutely usual—in the sense that it happens all the time, to every material object in the Universe. It's simply that h is so very small. Our ordinary methods of observation are too crude; they mask this underlying type of indeterminacy. And it's this that misleads people into thinking that position and velocity are in themselves definite quantities. They recognize that in purely practical terms you're never going to be able to *determine* what those values of position and velocity are—not to *infinite* precision—but this they put down to nothing more than the clum-siness of their measuring techniques. But in truth, both quantities are *fundamentally* indefinite to some extent.

'Actually it is possible to alter the *balance* of uncertainties. For example, you might want to concentrate on improving the accuracy of your determination of position. OK you can do that, but the price you have to pay is an increase in the uncertainty of the velocity. Alternatively you can go for precision of velocity, but then you have to sacrifice precision of position. Planck's constant governs the relation between these two uncertainties.'

> 对于"测不准原理"（这里正是对其的形象说明），那可真是量子惹的祸。

'I'm not altogether sure…,' began Mr Tompkins.

'Oh it's quite simple really,' continued the professor. 'Look here, I am going to put definite limits on the position of this ball.'

The fuzzy-looking ball he spoke of was lazily rolling over the table. He reached across and trapped it inside the wooden triangle the players use for setting up the balls at the start of a game. Immediately the ball seemed to go beserk. The whole of the inside of the triangle became filled up with a blur of ivory.

> 这个对测不准原理的说明会更准确些，也同样形象化。

'You see!' said the professor, 'I have now defined the position of the ball to the extent of the dimensions of the triangle. Previously all we could say for certain was that it was on the table—somewhere. But look what it's done

8 Quantum Snooker

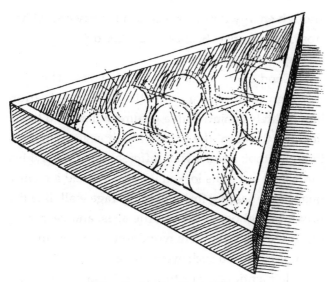

Ball confined to lie within a triangle

to the velocity. The uncertainty in the velocity has shot up.'

'Can't, you stop it rushing about like that?' asked Mr Tompkins.

'No—it's physically impossible. Any object in an enclosed space has to possess. a certain motion-we physicists call it *zero-point motion*. It's impossible for it to stay still. If it *did* stay still then we would know for certain what its velocity was; it would be zero. But we are not allowed to know the velocity if we have a pretty good fix on its position-as we do here with the ball confined to lie within the boundaries of the triangle.'

While Mr Tompkins was watching the ball dashing to and fro in its enclosure—like a tiger in a cage—something very odd happened. The ball got out! It was now on the *outside* of the triangle, rolling towards a distant corner of the table. But how? It wasn't that it jumped over the wall of the triangle; instead it had sort of 'leaked' through the barrier.

'Hah!' exclaimed the professor excitedly, 'Did you see that? One of the most interesting consequences of quantum theory: It is impossible to hold anything inside an enclosure indefinitely—provided there is

零点运动，这可是对于"世界是运动的"这一命题最彻底的诠释。

这就是量子理论中心"隧道效应"了。

在量子力学中，人们甚至不能，也不应该想，在"隧道效应"中"漏出来"的球是以什么方式穿过木框的，因为那种想像的方式就是非量子理论的。

enough energy for the object to run away once it has crossed the barrier. Sooner or later the object will "leak through" and get away.'

'Good grief!' declared Mr Tompkins. 'Then I'll never go to the zoo again.' His vivid imagination immediately conjured up a picture of lions and tigers 'leaking through' the walls of their cages. Then his thoughts took a somewhat different turn: What if his car leaked out of its locked garage? He had a mental image of it passing through the garage wall, like the proverbial ghost of the middle ages, and careering off down the street. He wondered whether his car insurance covered such eventualities.

He mentioned this to the professor, and asked, 'How long would I have to wait for that to happen?'

After making some rapid calculations in his head, the professor came back with: 'It will take about 1,000,000,000 …000,000 years.'

Even though Mr Tompkins was accustomed to large numbers in the bank's accounts, he lost count of the number of noughts in the professor's answer. It was, however, a reassuringly long period of time—enough for him not to be unduly worried.

'But tell me,' he said. 'In the ordinary world—in the absence of balls like these—how can such things be observed if it takes so long to happen?'

'Good question. There's no point hanging around hoping to see ordinary, everyday objects perform these feats. No, the point is the effects of the quantum laws only really become noticeable when you're dealing with very small masses such as atoms or electrons. For such tiny particles, the quantum effects are so large that ordinary mechanics becomes quite inapplicable. A collision between two atoms, say, would look exactly like a collision between two of these "quantum-elephantistic" balls. Not only that, but the motion of electrons within an atom resembles very closely the "zero-point motion" of the ball when it was inside the wooden triangle.'

> 关于汽车从车库中"漏"出来之可能性的例子，是一个很好的例子，可以说明宏观世界与微观世界在理论解释上的统一性，如此之小的可能性，也就等于是"不可能"——对于宏观世界的不可能。

> 因此，我们大可不必因微观理论而为我们"安全"的宏观世界的生活而担心。

8 Quantum Snooker

What if his car leaked out of its locked garage?

'And do the electrons escape from their atoms very often?' asked Mr Tompkins.

'No, no,' responded the professor hurriedly. 'No, that doesn't happen at all. You must remember what I said about the object having enough energy to get away once it has leaked through the barrier. An electron is held in an atom by the force of attraction between the negative electric charge it carries and the positive charge on the protons in the nucleus. The electron does not have enough energy to escape this pull, so it cannot get away. No, if you want to see leakage, then I suggest examining the nucleus of the atom. To some extent a nucleus can behave as though it's made up of alpha particles.'

'Alpha particle?'

'That's the name given historically to the nucleus of a helium atom. It consists of two neutrons and two

> 这段梦也做得有些抽象和理论化了。

111

protons. It is exceptionally tightly bound; the four particles can "fit together" in a very efficient manner. Anyway, as I was saying, because alpha particles are so tightly bound, heavy nuclei can in some circumstances behave as though they were a collection of alpha particles-rather than individual neutrons and protons. Although the alphas are moving about within the overall volume of the nucleus, they are constrained to stay within that volume by the short-range attractive forces that bind nuclear particles together. At least, they stay together normally, but every so often, one of the alphas escapes. It gets out beyond the range of the attractive nuclear force that had been constraining it. In fact, now it is subject only to the long range repulsive force between its positive electric charge and that on the rest of the nucleus it has left behind. So now the alpha is propelled away. It's a form of *radioactive nuclear decay*. So, as you see, this is quite analogous to your car in its garage—only the alpha escapes more quickly!'

> 梦做得太难时，就该醒了。

At this point, Mr Tompkins felt a strange sensation in his arm. It had begun to shake. He heard a woman's hushed voice saying 'Shh!'

He awoke to find a lady sitting next to him on the lecture theatre bench. She was gently tapping him on the arm. She smiled sympathetically and whispered. 'You were beginning to snore.'

> 由此我们看到，虽然教授尽力要通俗，但无论如何演讲还是要比做梦抽象难懂，但由此我们也正可以比较一下，因为这里正是用理论来解释梦中的那些"现象"。

Mr Tompkins pulled himself together and silently mouthed the words 'thank you' to her. He wondered how much of the lecture he had missed. Perhaps even in his sleep he had been unconsciously tuned in. He remembered hearing a report once of someone who was supposed to have learned a foreign language by going to sleep with headphones on. Anyway, the professor was still in full flow…

Let us now return to our experimenter and try to get the mathematical form for the limitations imposed by quantum conditions. We have already seen that whatever method of observation is used, there is always a conflict between the estimate of position and that of velocity of the moving object. In the optical method, the collision between the object and the photon from the illuminating source will, because of the law of conservation of momentum, introduce an uncertainty in the momentum of the particle comparable with the momentum of the photon used. Thus, using equation (15), we can write for the uncertainty of momentum of the particle

$$\Delta p_{particle} \cong h/\lambda \qquad (16)$$

Remembering that the uncertainty of position of the particle is given by the wave length (ie. $\Delta q \cong \lambda$) we deduce:

$$\Delta p_{particle} \times \Delta q_{particle} \cong h \qquad (17)$$

In the mechanical method of observation using the "bells", the momentum of the moving particle will be made uncertain by the amount taken by the bell clapper. Because the clapper is confined within the bell, its momentum must be such as to correspond with a wave length comparable to the dimensions of the bell, i. Thus, using equation (15), $\Delta p_{particle} \cong h/l$. Recalling that in this case the uncertainty of position is given by the size of the bell (i.e. $\Delta q \cong l$), we come again to the same equation (17). This universal relationship between the two uncertainties, involving as it does Planck's constant, was first formulated by the German physicist Werner Heisenberg. Hence, equation (17) is known as *the Heisenberg uncertainty relationship*. From this it becomes immediately clear that the better one defines the position, the more indefinite the momentum (or velocity) becomes, and vice versa. Remembering that momentum is the product of the mass of the moving particle and its velocity, we can write

$$\Delta v_{particle} \times \Delta q_{particle} \cong h/m_{particle} \qquad (18)$$

For bodies which we usually handle, these uncertainties are exceedingly small. For a light particle of dust, with a mass of 0.000, 000, 1g, both position and velocity can be measured with an accuracy of 0.000, 000, 01%! However, for an electron (with a mass of 10^{-30}, kg) the product $\Delta v \Delta q$ should be of the order of 10^{-4} m²/s. Inside an atom, the velocity of an electron should be less than 10^6 m/s, otherwise it will escape. So its velocity needs to be defined to a precision within that limiting velocity. Using equation(17), this gives 10^{-10} m for the uncertainty of position, i.e. we would expect that this would represent the total dimensions of an atom. And indeed, that is what we find to be the case in practice. Here we begin to glimpse the power and usefulness of Heisenberg's uncertainty relationship. Merely from a knowledge of the strength of the forces within the atom (and hence the maximum velocities allowed to the electrons), we are able to arrive at an estimate of the size of atoms!

In this lecture I have tried to show you a picture of the radical change that our classical ideas of motion have had to undergo. The elegant and sharply defined classical notions are gone, and you might well be wondering how physicists manage to keep afloat on this ocean of uncertainty. It cannot be a function of an introductory lecture like this to provide you with the full mathematical rigour of quantum mechanics. But for those of you interested, let me give you the flavour of it.

It is clear that if we cannot in general define the position of a material particle by a mathematical point, and its trajectory by a mathematical line, we have to use other mathematical methods of description. This in fact entails the use of continuous functions (such as are used in hydrodynamics). Such functions will allow us to define the 'density of presence' of the object as it 'spreads out' in space.

I should perhaps warn you against the erroneous idea that the function describing the 'density of presence' has a physical reality in our ordinary three-

> 这里已经在用理论讲梦中形象比喻的不严格性了。

dimensional space. In fact, if we describe the behaviour of, say, two particles, we must answer the question concerning the presence of our first particle in one place and the simultaneous presence of our second particle in some other place. To do this we have to use a function of six variables (the coordinates of the two particles), and these cannot be 'localised' in three-dimensional space. For more complex systems, functions of still larger numbers of variables must be used. In this sense, the quantum mechanical *wave function* is analogous to the 'potential function' of a system of particles in classical mechanics or to the 'entropy' of a system in statistical mechanics. It only *describes* the motion, and helps us to predict the relative probabilities of various possible outcomes of our next observation of the object. For example, suppose we have an electron beam being diffracted as it passes through slits in a barrier, before finally striking a distant screen where its arrival is recorded. The wave function for this physical set-up will allow us to calculate the relative probabilities for the electrons arriving at different locations on the screen—their arrival being in the form of localised quanta or particles.

The Austrian physicist Erwin Schrodinger was the first to write the equation defining the behaviour of the wave function, Ψ, of a material particle. I am not going to enter here into the mathematical derivation of his fundamental equation, but 1 will draw your attention to the requirements which lead to it, the most important of these being a very unusual one: The equation must be written in such a way that the function describing the motion of material particles should show all the characteristics of a *wave*.

Thus, the behaviour of our Ψ function is not analogous to (let us say) the passage of heat through a wall heated on one side, but rather to the movement of a mechanical deformation (a sound wave) through the same wall. Mathematically, it requires a definite rather restricted form of equation. This fundamental condition, together with the additional requirement that our equations. should go over into the equations of classical mechanics when applied to particles of large mass for which quantum effects should become negligible, practically reduces the problem of finding the equation to a purely mathematical exercise.

If you are interested in how the equation looks in its final form, here it is:

$$\nabla^2 \Psi + \frac{4\pi m i}{h} \Psi - \frac{8\pi^2 m}{h} U\Psi = 0 \qquad (19)$$

> 有人说，科普著作中，一个数学公式会吓走一半读者，可作者在这里没少用数学公式，这也是不得已，但对大多数读者，也许并不需要太深究，体会一下数学语言的风格就可以了。

In this equation, the particles have mass, m, and the function U represents the potential of the forces acting on the particles. The equation gives the solution for the motion of the particles, given the particular distribution of forces. The application of *Schrödinger's wave equation* has allowed physicists to develop the most complete and logically consistent picture of all phenomena taking place in the sub-atomic world.

Before I end I suppose I ought just to mention a word or two about matrices. If you have already read quite a bit about quantum physics, you may have come across this very different approach to the subject. I must confess that personally I rather dislike these matrices, and prefer to do without them. But, for completeness I ought at least to mention them.

The motion of a particle or of a complex mechanical system is always described, as you have seen, by certain continuous wave functions. These functions are often rather complicated and can be represented as being composed of a number of simpler oscillations, the so-called 'proper functions,' much in the way that a complicated sound can be made up from a number of simple harmonic notes. One can describe the whole complex motion by giving the amplitudes of its different components. Since the number of components (overtones) is infinite we must write infinite tables of amplitudes in a form:

$$
\begin{array}{cccc}
q_{11} & q_{12} & q_{13} & \cdots \\
q_{21} & q_{22} & q_{23} & \cdots \\
q_{31} & q_{32} & q_{33} & \cdots \\
\cdots & \cdots & \cdots & \cdots
\end{array}
\qquad (20)
$$

Such a table, which is subject to comparatively simple rules of mathematical operations, is called a 'matrix'. Some theoretical physicists prefer to operate with matrices instead of dealing with the wave functions themselves. Thus, the 'matrix mechanics', as they sometimes call it, is just a mathematical modification of the ordinary 'wave mechanics'.

I am particularly sorry that time does not permit me to describe to you the further progress of quantum theory in its relation to the theory of relativity. This development, due mainly to the work of the British physicist Paul Dirac, brings in a number of very interesting points and has also led to some extremely important experimental discoveries. I may be able to return at some other time to these problems, but bere for the present I must stop.

> 这里在讲"矩阵"了。不过对量子理论，究竟用"矩阵"的方式，还是用波动力学的方式来描述，这更多地是科学史上的问题，现在波动方程的表示明显地为人们更爱采用。

9 The Quantum Safari
量子丛林

> 看着说的像真的似的，其实还是一个梦。

Beep…beep…beep

Mr Tompkins pulled himself up, reached out from under the bedclothes and banged the top of the alarm clock. Thoughts of Monday morning and work began to seep into his consiousness. Slumping down  once more, he began his customary final ten minutes snooze before the insistent noise was due to recommence…

'Hey! Come on! It's time to get up. We've got a plane to catch, remember.' It was the professor; he was standing by the side of the bed, holding a large suitcase.

'What…what's that?' mumbled a flustered Mr Tompkins as he sat up rubbing his eyes.

'We're going on safari. Don't tell me you've forgotten!'

'Safari!?'

'Yes of course. We're off to the quantum jungle. Very helpful the landlord of that pubtold me where the ivory for his snooker balls came from.'

'Ivory?! But we're not supposed to go looking for ivory these days…'

Brushing aside Mr Tompkins' protest, the professor rummaged in the side pocket of the case.

'Ha!Here it is,' he declared, pulling out a map.'Yes, look. I've marked the region in red. See? Everything within that area is subject to quanturm laws with a very large value of Planck's constant. We're off to investigate.'

> "量子丛林"的梦主要把宏观世界的量子效应放大，反过来想，把人变小到微观世界，是不是也类似呢？

The journey was nothing remarkable, and Mr Tompkins scarcely noted the time until the plane touched down at their destination in some distant country. According to the professor,this was the nearest

9 The Quantum Safari

populated place to the mysterious quantum region.

'We shall be needing a guide,' he said. But recruiting one turned out to be difficult. The locals were clearly wary about going to the quantum jungle, and normally never went near the place. But eventually a brash, dare-devil lad, taunting his friends for their cowardice, volunteered to take the two visitors.

> 在真实世界旅行还经常需要导游呢，何况在量子世界里？

First stop was the market to pick up supplies.

'You'll have to rent us an elephant to ride,' the boy announced.

> 不知给养里有无食品，也不知食品的量子效应是什么。

Mr Tompkins took one look at the huge animal, and was immediately filled with alarm. He was expected to mount *that*! 'Look, I'd rather not,' he declared. 'I've never done this sort of thing before. I really couldn't. A horse, maybe. But not that.' Just then he noticed another trader selling donkeys. He brightened. 'How about one of them? I reckon that's more my size.'

The boy laughed derisively. 'Take a donkey to the quantum jungle? You must be joking. That would be like riding a bucking bronco. You'd be thrown off in no time(assuming the donkey didn't leak through your legs before that).'

'Ah yes ,'murmured the professor. 'I begin to see. The lad's actually making a lot of sense.'

'He is?' said Mr Tompkins. 'I reckon he's in cahoots with the elephant salesman. They're ripping us off—making us buy something we don't need.'

'But we *do* need an elephant,' replied the professor. 'We can't ride an animal that's going to spread out all over the place—like those snooker balls. We need to be attached to something heavy. That way the momentum will be high, even if it's going slow, and that in turn means the wavelength will be tiny. I told you some time ago that all the uncertainty in position and velocity depends on the mass; the larger the mass, the smaller the uncertainty. That is why the quantum laws have not been observed in the ordinary world even for objects as light as particles of dust. Electrons, atoms and molecules, yes. But not

> 由此我们可以设想，假如在我们生活的世界里，普朗克常数很大，又会是什么样的情景。

ordinary sized objects. In the quantum jungle, on the other hand, Planck's constant is large. But even there, it's still not large enough to produce striking effects in the behaviour of such a heavy animal as an elephant. The uncertainty of the position of a quantum elephant can be noticed only by close inspection. One might expect its outline to be slightly fuzzy, but nothing more. In the course of time, this uncertainty will increase very slowly. That in fact is probably the origin of the local legend that very old elephants from the quantum jungle develop long fur.'

After some haggling, the professor agreed a price, and he and Mr Tompkins mounted the elephant, climbing into the basket fastened to the animal's back. With the young guide taking up his position on the elephant's neck, they started towards the mysterious jungle.

It took about an hour to reach its outskirts. As they entered the forest, Mr Tompkins noted that the leaves in the trees were rustling, and yet there did not appear to be a wind. He asked the professor why this was so.

'Oh, that's because we're looking at them,' was the reply.

'Looking at them! What's that got to do with it?' exclaimed Mr Tompkins. 'Are they so shy?'

> 这就是量子世界中观察者对被观察对象之干扰的形象描述，应该庆幸的是在我们的世界里普朗克常数如此之小，才使我们免去了被别人"看"伤的危险。

'I would hardly put it like that,' smiled the professor.'The point is that in making any observation you can't help disturbing whatever it is you are looking at. The photons of sunlight here obviously pack a bigger punch than the ones we are used to back home. With Planck's constant being that much bigger now, we must expect to find it a pretty rough world. No gentle action is possible here. If a person here tried to pat a dog, it would either not feel anything at all, or its neck might be broken by the first quantum of caress.'

As they ambled along through the trees, Mr Tompkins got to thinking. 'What if nobody is looking?' he asked. 'Would everything behave properly then? I mean, would those leaves behave in the way we are accustomed

9 The Quantum Safari

to think?'

'Who can say?' mused the professor. 'When nobody is looking, who can know how they behave?'

'You're saying that is more a philosophical question than a scientific one?'

'You can call it philosophy if you like. But it might simply be a meaningless question. One thing is clear, in *science* at least, it is a *fundamental* principle that one tries *never to speak about the things you cannot experimentally test*. All modern physical theory is based on this principle. In philosophy it might be different. Some philosophers might try to go beyond that. For example, the German philosopher Immanuel Kant spent quite a lot of time reflecting about the properties of objects not as they "appear to us", but as they "are in themselves". For the modern physicist only the so-called "observables" — the results of measurements, such as position and momentum—only these have any significance. All of modern physics is based on their mutual relation⋯'

> 在终极的意义上，物理学与哲学是不可分的。

At this moment there was a sudden buzzing noise. They looked up and momentarily caught sight of a large black flying insect. About twice the size of a horse fly, it looked exceedingly vicious. The boy guide yelled a warning to them to keep their heads down. He produced a fly swat, and immediately started thrashing out at the attacking insect. The insect became a blur, and the blur, in its turn, developed into an indistinct cloud which enveloped the elephant and its riders. The boy was now vigorously swatting in all directions, but mostly at the region where the cloud was densest.

> 密度最大的地方，也就是昆虫出现的概率最大的地方。

THWACK! He succeeded in making contact. The cloud instantly disappeared, and the dead body of the insect could be seen hurtling away, describing an arc in the air, and landing somewhere among the dense undergrowth.

'Well done!' exclaimed the professor. The boy beamed triumphantly.

'I'm not sure I quite understood what that was all

about···' murmured Mr Tompkins.

'Oh nothing to it really,' replied the professor. 'The insect is very light. After our first sighting of it, its position rapidly became more and more uncertain with time. Eventually we were surrounded by an "insect probability cloud"—much in the same way as an atomic nucleus is surrounded by an "electron probability cloud". By the time that had happened, we no longer knew for certain where to find the insect. Except that where the probability cloud is densest, that is where there is the better chance of finding it. Didn't you see how the lad was preferentially swatting at the denser parts of the cloud?' That was the correct strategy. It increased the probability of the interaction between the swatter and the insect. In the quantum world, you see, one cannot aim precisely and be sure of a hit.

As they resumed their journey, he continued. 'That's exactly what we find in our world at home but on a much smaller scale. The behaviour of the electron round the atomic nucleus is in many respects analogous to the behaviour of the insect which seemed to be all round the elephant. With atomic electrons you have no more certainty of being able to hit it with a photon, say, as the lad had of hitting the insect. It's all down to probabilities—playing the odds. You shine a beam of light on the atom and most of the photons will miss; they pass through without having any effect at all. You just hope one of the photons will score a bullseye.'

'Sounds like the poor dog which cannot be patted in the quantum world without being killed,' concluded Mr Tompkins.

Just then they emerged from the forest and found themselves on a high plateau overlooking open country. The plain laid out below them was divided in two by a dense line of trees, hugging the banks of a dried-up river bed, and stretching away from them into the distance.

'Look! gazelles, and lots of them!' whispered the professor excitedly, pointing to a herd quietly grazing

> 用母狮和羚羊来形象地表现宏观尺度上的"双缝干涉"现象,实在是很有想像力。

over to the right.

But Mr Tompkins's attention had been drawn to what lay on the opposite side of the tree line. He had seen a group of lionesses. Then, a short way off, he spotted another group, and another, and another…The groups of lionesses were strung out along a straight line running parallel to the trees. Moreover, the groups were exactly equally spaced from each other. How very odd, he thought. It reminded him of the scene that greeted him every morning, Monday to Friday, on the railway station platform at home. Through long experience, regular commuters on the 7.05 a. m. knew exactly where the doors of the train were likely to be when the train pulled in. Unless you were situated directly opposite a door when it opened, you had no chance of getting a seat. That is why the old hands like Mr Tompkins were to be found huddled together in small groups at regular intervals along the platform.

The lionesses were all looking expectantly towards two narrow gaps in the tree line. But before Mr Tompkins could ask what was going on, there was a sudden commotion over at the far right hand side. A solitary lioness had suddenly emerged out into the open from where she had been concealed. The gazelles caught sight of her and immediately took fright. They fled in a headlong charge towards the two gaps in the trees.

> 在量子效应下,羚羊的可怜命运!

As they emerged the other side, a most bizarre thing happened. Instead of staying together as a herd, or scattering in all directions, they took off in separate columns—each column *heading straight for one of the groups of waiting lionesses*. On arrival, the kamikaze gazelles were duly set upon, and eaten.

Mr Tompkins was dumb-struck, 'That doesn't make sense,' he cried.

'Oh but it *does*,' murmured the professor. 'It most certainly does. How absolutely fascinating. Young's double slit.'

'Whose double what', moaned Mr Tompkins.

9   The Quantum Safari

'Oh, sorry. More jargon, I'm afraid. The point is that there is an experiment where you shine a beam onto two slits in a barrier. If the beam were made of particles—like paint being sprayed out of a can—you'd expect to get two beams coming out the other side, one corresponding to each slit. But if the beam is made of waves, each slit acts as a source of waves. They spread out on the far side, overlapping with each other. The humps and troughs of the two lots of waves get mixed up with each other and interfere with each other. In some directions the wave trains are out of step; the humps from one coincide with the troughs of the other; they cancel each other out, so nothing happens in those directions. We call that *destructive interference*. In other directions, we get the opposite: the wave trains are in step, the humps of one coincide with the humps of the other, and similarly the troughs coincide; they reinforce each other and you get a particularly large transmitted wave in those directions. That is what we call *constructive interference*.'

'So you're saying that you get separated beams on the other side of the slits—where you get constructive interference—and nothing in between where it's destructive?' asked Mr Tompkins.

杨氏双缝实验是物理学史上最重要的实验之一，它证明了光的波动性。

'That's right. And you are not confined to just two beams. You can get lots of them, equally separated from each other. The angle they come off at depends on the wavelength of the initial beam and on the distance separating the two slits. The fact that you get more than two transmitted beams proves that you are dealing with waves rather than particles. It's called "Young's double slit experiment" because that is how the physicist Young was able to demonstrate that light beams are made up of waves. Now, in this version here,' the professor gestured to the scene of carnage below them, 'you have a demonstration that gazelles also have a wave behaviour.'

'But I still don't quite understand,' the puzzled Mr Tompkins persisted. 'Why did the gazelles commit suicide like that?'

这幅图画得很巧，没有画出羚羊究竟是具体通过哪个"缝"跑过去的。在量子力学对双缝干涉实验的解释中，说从哪个"缝"穿过是没有意义的！

IT REMINDED HIM OF...THE RAILWAY STATION PLATFORM

'They had no alternative. The interference pattern determined where they were likely to end up. For any particular gazelle there was no way of telling in which direction it would go on emerging from the two gaps in the trees. All you can say in advance is that the probabilities will be high in certain directions and poor in others. The gazelles have simply to go through the gaps and see what happens. Unfortunately for them, the lionesses are experienced hunters. They know how much the average gazelle weighs and how fast it can run. So that determines the momentum and hence the wavelength of

9 The Quantum Safari

the gazelle beam. They also know the distance between the gaps in the trees, and that way can work out where they should wait for their meal to be delivered.'

'You mean those lionesses are good at mathematics!?' exclaimed Mr Tompkins incredulously.

The professor laughed. 'No. I doubt that. No more than a child has to be good at calculating parabolic trajectories to know how to catch a ball. It's probably an instinctive judgement they're making.'

As they watched, the lioness that had originally scared the herd to take flight herself joined one of the groups of lionesses for her share of the meal.

'Nice touch that,' observed the professor. 'Did you note how slowly she ambled up to the gaps in the trees. She was obviously compensating for having a greater mass than the typical gazelle. By moving more slowly she ended up with the same wavelength. That way she ensured that she herself would get diffracted into one of the directions followed by the gazelles—and so would end up with one of the meals. An evolutionary biologist would have a field day studying the kinds of behaviour that have been selected for in this environment…'

He was interrupted by a high-pitched buzzing sound.

'Look out!' cried the guide. 'Another insect is about to attack.'

Mr Tompkins hurriedly crouched down low, drawing his coat over his head for added protection. Except that it wasn't his coat, but a bed sheet. And it wasn't the sound of an attacking quantum insect, but the beeping of the alarm clock.

> 是啊，生物的不可思议之处也在此，哪怕是最优秀的篮球运动员在投篮前，也不会真的用数学计算投篮的用力和篮球的轨道，由此来看，用数学计算倒是一件很笨的事。

> 梦醒时分。

10 Maxwell's Demon

麦克斯韦妖

Over the ensuing months, Mr Tompkins and Maud visited art galleries together and argued the merits or otherwise of the exhibitions they had seen. He tried as best he could to introduce her to the mysteries of the quantum physics he had recently learned. With his head for figures he was also proving invaluable to her when it came to lending a hand with the business side of her involvement with dealers and art gallery owners.

> 因为科学而成就一段姻缘。

In due course, he plucked up courage to ask her to marry him, and was delighted when she accepted. They decided to set up home at Norton Farm; that way she would not have to surrender her studio.

One Saturday morning, they were expecting her father round for lunch. Maud was sitting on the sofa reading the latest *New Scientist*. Mr Tompkins was at the dining table trying to sort out her tax return. Sifting through the piles of receipts for art materials, he commented, 'I don't see myself being able to retire early and live off my wife's earnings—not yet anyway.'

> 可艺术不能当饭吃，于是……

'And I don't see the two of us living off yours either,' she replied without looking up.

Mr Tompkins sighed, gathered up the papers and put them back in a box file. He picked up the newspaper and joined Maud on the sofa. Leafing through the colour supplement, his attention was drawn to an article about gambling.

'Hey,' he said after a moment. 'I reckon this is the answer. A fool proof betting system'.

10 Maxwell's Demon

'Oh, yes,' murmured Maud absently, still reading. 'Who says?'

'It says it here.'

'Oh, it's in the paper, then it must be true.' she said sceptically.

'No, seriously. Listen to this. You bet on the first horse so as to win £1, say. If you win, fine. You put that £1 in the bank.'

'And if you lose?'

'If you lose, you bet on a second horse, but now you raise the stake so that if you win, you get back what you lost on the first race, plus £1. That way you get your pound to put in the bank, and you've not lost anything. If you lose the second time, then on the third race you raise the stake so that you recover your losses on the first two races, plus you gain the pound. It's simple. That way, it doesn't matter how often you lose, in the end you're bound to get your money back from the previous races—that was just temporary—*and* you make a profit of £1 into the bargain.'

'Well a pound's not much,' said Maud, still not convinced.

像这种初听上去似乎有道理的赌法，不妨仔细思考一下它隐藏的陷阱。

You're bound to make a profit

'But that's just for starters,' said Mr Tompkins excitedly. 'It goes on here to say that having put the £1 winnings in the bank, you don't touch it. Instead you go on to repeat the whole thing all over again. You back a horse to win £1; if you lose you increase the stake so as to cover the loss, plus make a profit of £1. You carry on like this until you eventually win again—and that gives you a second pound to put in the bank. Now you've got £2, and so on. £3, £4, *ad infinitum*. How about that!' he ended triumphantly.

'Well, I don't know,' she replied uncertainly, 'Father has always said there can't be such a thing as a sure-fire gambling system.'

> 这位物理学家父亲的名言，倒是不变的真理。

'Oh no? Where's the flaw?' he demanded. 'Tell you what: I'll *prove* it. I'll put it into practice right away.' With that, he turned to the racing pages of the sports section, shut his eyes and made a stab with his finger. 'Demon's Delight, the 2.30 at Haydock. That'll do as well as any. OK. I'll pop down to the betting shop right this very minute.'

He rose, put on his jacket and made for the door. But before he could reach it, the door bell rang; it was Maud's father.

'Oh, you off somewhere?' the professor asked.

Mr Tompkins explained what he was doing.

'Oh, I see,' was the noncommittal reply. 'That old chestnut.' The professor passed Mr Tompkins in the passage and went to greet his daughter. Being a nice warm day, they then made their way out to the seats on the patio.

'A fool-proof gambling system?' he muttered. 'The number of times I've heard *that*.'

> 与其说汤普金斯相信了这种赌法的"证明"，倒不如说是他想赢钱的欲望在帮助他幻想会有一种不输的赌法。

'I know it sounds unlikely,' admitted Mr Tompkins, as he followed the professor. 'But this one is different. You're guaranteed not to lose. You're *bound* to win. This one cannot miss.'

'Can't it?' said the professor with a smile. 'Well, let's see, shall we.' After a short inspection of the article, he went on, 'The distinguishing feature of this system is that the rule governing the amount of your bets calls for you to raise your bet after each loss. If you should win and lose alternately and with complete regularity, your capital would oscillate up and down, each increase being slightly larger than the previous decrease. In such a case you would, of course, gradually increase your capital over time, possibly becoming a millionaire in due course.'

'That's what I said.'

'But as you are no doubt aware, such regularity does not occur,' continued the professor. 'As a matter of fact, the probability of such a regularly alternating series is just as small as the probability of an equal number of straight wins. So we must see what happens if you have a sequence of several successive wins or losses.

'If you get what gamblers call a streak of luck, you will make a succession of wins. But your total winnings, at £1 a time, will not be very high. On the other hand, a bad streak will rapidly land you in deep trouble. You will find that the rate at which you have to increase your bet to cover previous losses will quickly clean you out and throw you out of the game. For example, if the odds are even(you bet £1 to win £1), after five successive losses you will have to bet £32 next time to cover those losses and make a profit of £1; ten straight losses and the stake becomes £1,024, fifteen losses and you are now having to bet 32,768—all in order to win £1! A graph representing the variations in your capital will consist of several slowly rising portions interrupted by very sharp drops. At the beginning of the game, it is likely that you will get onto the long, slowly rising part of the curve and will enjoy for a while the pleasant feeling of watching your money slowly but surely increasing. However, if you go on long enough, in the hope of larger and larger profits, you will come unexpectedly to the sharp drop which will be deep enough to make you bet—and lose—your last penny.

> 这才是赌博者长赌必输的硬道理!

> 幸好人生毕竟不是赌场——尽管有时也有点赌的意味。

'The important thing is this: no purse is infinite. Any gambler embarking on this scheme will have limited funds. They may be great, but they are necessarily *limited*. That being so, there *must* come a time when, by the law of averages, a bad running streak will produce losses sufficient to wipe out those funds. In a quite general way, with this or any other similar system, the probability that you will double your initial funds from winning is equal to that of having your funds wiped out. In other words, the chances of finally winning are exactly the same as if you put all your money on the toss of a coin—double or quits. All that such a system can do is to prolong the game and give you more fun (or agony) for the money.

'Of course I have been assuming all along that the bookmaker is not taking a cut, which is not true, and means things are even worse than I have described. No, the only person guaranteed to end up happy and prosperous by your foolproof system is the bookmaker.'

> 这里的风险,几乎是确定的。

'So, you're saying there's no possible way of winning money without risking the slightly higher probability of losing it?' said Mr Tompkins dejectedly.

'Precisely,' said the professor. 'What is more, what I have said applies not only to such comparatively unimportant problems as games of chance, but to a great variety of physical phenomena which, at first sight, seem to have nothing to do with the laws of probability. For that matter, if you could devise a system for beating the laws of chance, there are much more exciting things than winning money one could do with it. One could build cars that ran without petrol, factories that could be operated without coal or oil, and plenty of other fantastic things.'

> 从关于赌博的数学,自然轻松地就过渡到了物理,而且还是关于"永动机"的重要话题。

'Really?' asked Mr Tompkins, beginning to take an interest. He sat down on the sofa once more. 'I've read about machines like that. Perpetual motion machines, right? But I thought you couldn't have such a thing—machines running without fuel. You can't manufacture

10 Maxwell's Demon

energy out of nothing.'

'Quite right, my boy,' agreed the professor, pleased to think that he had been able to deflect his son-in-law's attention away from hare-brained gambling systems, and back to his own favourite topic: physics. This kind of perpetual motion. "perpetual motion machines of the first type" as they are called, cannot exist because they would be contrary to the law of the Conservation of Energy. However the fuelless machines I have in mind are of a rather different type. They are usually known as "perpetual motion machines of the second type". They are not designed to create energy out of nothing, but to *extract* energy—already existing energy from surrounding heat reservoirs in the earth, sea or air. For instance, you can imagine a steamship where the boilers produced steam by extracting heat from the surrounding water—instead of burning coal or oil. It would depend on you being able to force heat to flow away from something that was cold toward something that was hotter—which, of course, is the opposite of what heat normally does.'

> 要注意区分这两类永动机。

> 永动机的不可能，就像赌博不会包赢不输一样，世界上没那么好的便宜事！

'That sounds a great idea,' enthused Mr Tompkins. 'We could construct a system for pumping in sea-water, taking out its heat content, the heat goes to the boilers, and the rest—the blocks of ice—well, we could just throw those overboard. In fact, I seem to remember from school that when a gallon of cold water freezes into ice, it gives off enough heat to raise another gallon of cold water almost to boiling point. Right? So all we would have to do is pump a few gallons of seawater per minute, and we could easily collect enough heat to run a good-sized engine. You know, I reckon we're on to something.'

> 了不起的想法，未必就会是可实现的。

'Lunch is on the table,' Maud called out from the dining room. The two men, who had been so intent on their conversation that they had not even noticed that Maud had left them to prepare the meal, reentered the house themselves and joined her.

'Forget the bookmaker, Maud,' Mr Tompkins said as they sat down. 'Your Dad's onto a real sure-fire cert here!'

Having helped himself to vegetables, he paused and frowned. He turned to the professor, 'Except... if it's such a good idea, why hasn't someone thought of it before?'

The professor smiled. 'But they have. You see, for all practical purposes, such a perpetual motion machine—of the second type—would be just as good as the kind designed to create energy out of nothing. With engines like that to do the work, you'd never have to worry about fuel bills or conserving energy resources. The trouble is machines of that type are just as impossible as the first type.'

'But why?'

'The laws of probability,' replied the professor. 'The same laws as defeated your foolproof gambling system.'

> 数学中的概率定律，既决定了赌博的结果，也决定了永动机的不可能。

'Sorry? I don't see the connection. What have laws of probability got to do with it?'

'Well, heat processes are themselves subject to probability; they're very similar to gambling games—betting on horses, rolling dice, spinning the roulette wheel—that sort of thing. To expect heat to flow from something cold to something that is hotter... well, that would be like hoping that money will flow from the bookmaker's bank into your pocket. Or expecting the salt to sprinkle itself over my plate without someone giving it a helping hand.'

'Salt? What...?'

'Cyril,' chided Maud gently, nodding in the direction of the salt-cellar.

'Oh, I'm sorry,' he said apologetically, passing it to his father-in-law. 'Wasn't thinking.'

> 太数学化的话题显然不利于食欲和消化。

'How about changing the subject,' Maud suggested. 'At least until you've eaten.'

After lunch they decided to take their coffees outside. The professor asked Mr Tompkins if he might have a whisky. 'Just occasionally, my boy. Not used to

10 Maxwell's Demon

big lunches. Helps settle my stomach.'

Having settled into sun loungers, the professor whispered conspiratorially to Mr Tompkins, 'Do you think we can resume where we left off?'

Maud overheard and protested mildly, 'It is *Saturday*, you know. There should be a rule that no-one talks shop at the weekend.'

Ignoring her, they returned to the subject of probability.

'What do you know about heat?' asked the professor.

'A bit. But not much.'

'Right. Well it's nothing but the rapid irregular movement of atoms and molecules. You know, of course, that all material bodies are made of atoms? And some of the atoms stick together to form molecules?'

Mr Tompkins nodded.

'OK,' continued the professor. 'The more violent this molecular motion is, the warmer the body. As this molecular motion is quite irregular, it's subject to the laws of chance. It is easy to show that the most probable state of a system made up of a large number of particles will correspond to a more-or-less uniform distribution among all of them of the total available energy. If for any reason one particular part of the object gets heated—in other words, the molecules in this region are made to move faster—one would expect that, through a large number of accidental collisions, this excess energy would soon be distributed evenly among all the remaining particles.

'However, as the collisions are purely accidental, there is also the possibility that, merely by chance, a certain group of particles may collect the larger part of the available energy at the expense of the others.'

'You mean the temperature would rise? It would getter hotter in one place—and presumably colder in another?' ventured Mr Tompkins.

'Exactly. There would be a spontaneous concentration of thermal energy in one particular part of the object, and this would correspond to the flow of heat *against* the temperature gradient—from colder to hotter. This possibility is not excluded—at least not in principle. However, if one tries to calculate the relative probability of such a spontaneous heat concentration occurring, one gets such small numerical values that the phenomenon is for all practical purposes, impossible.'

'So have I got this right? You're saying that these perpetual motion machines of the second kind *could* conceivably work. They're not *absolutely* ruled out. But the chances of that happening are very slight—say, like throwing a couple of dice a hundred times and getting double 6 every time.'

> 在这里，物理学的道理和关于赌博的数学原理是统一的。

'Yes, that sort of thing. Except that the odds are much smaller than that,' said the professor. 'In fact, the probabilities of gambling successfully against nature are so slight that it is difficult to find words to describe them. For instance, I can work out the chances of all the air in the dining room collecting spontaneously under the table, leaving a vacuum everywhere else. The number of dice you would throw at one time would be equivalent to the number of air molecules in the room, so I must know how many there are. One cubic centimetre of air at atmospheric pressure, I remember, contains about 10^{20} molecules (1 followed by 20 noughts, yes?). So the air molecules in the whole room must total some 10^{27}. The space under the table is about... let's say one per cent of the volume of the room. That means the chances of any given molecule being under the table and not somewhere else are one in a hundred. So, to work out the chances of all of them being under the table at once, I must multiply one hundredth by one hundredth and so on, for each molecule in the room. And that gives one chance in 10^{54}.'

> 与现实相对应，我们通常认为那些概率很小的事不会出现，但从理论上讲，其实它们只是不大可能出现，而不是绝对不可能出现。

'Phew!' exclaimed Mr Tompkins, 'You'd have to be a pretty hardened gambler to bet on those odds!'

'Yes,' agreed the professor. 'You can take it from me you're not likely to suffocate because all the air lands up under the table. Nor for that matter, that the top half of the coffee in your cup will boil away while the bottom half becomes a block of ice.'

They laughed.

'But there is still a *chance* of the unusual happening,' insisted Mr Tompkins. 'Isn't there?'

'Yes, of course there is. It's not completely beyond the bounds of possibility for that flower pot over there to suddenly jump up into the air off the patio because the vibrations of the molecules in the ground accidentally received thermal velocities in the same upwards direction all at the same time.'

'Why that very thing happened only yesterday,' chimed in Maud. 'Remember, Cyril, how you were backing the car, and the dustbin...'

'All right, all right,' interrupted Mr Tompkins.

'What... what was that?' enquired the professor.

'Nothing nothing,' said Mr Tompkins hastily.

The professor chuckled. 'Well, whatever happened to the dustbin, I doubt you can lay the blame on Maxwell's Demon.'

'Maxwell's Demon? What do you mean?'

'Clerk Maxwell. A prominent physicist. He introduced the idea of a statistical demon. It was just a bit of fun. He used it to help explain what we've been talking about. Maxwell's Demon was supposed to be a very nimble fellow, capable of observing each individual molecule and changing its direction of motion in any way he wanted. If there really were such a demon, he would be able to deflect the motion of all the fast molecules so that they would go in one particular direction, and the slow ones he would deflect in the opposite direction. That way he could get heat to flow against the temperature gradient. It would be a way of going against the second law of thermodynamics: *the principle of increasing entropy.*'

麦克斯韦妖只是一种科学家在头脑中构想的东西。但这种理想化的构想却是科学家们有时会采用的推理论证方法之一。

'Entropy? What's that?'

'Oh, that's the term used to describe the degree of disorder of molecular motion in any given physical body or system of bodies. For instance, all the air molecules being under the dining room table and none anywhere else

"熵",是物理学中非常重要的基本概念,其意义也远远超出了物理学的范围。

in the room would be a very orderly arrangement. Having them Scattered all about the room willy-nilly would be very disordered. Or take the molecules in the surface of this patio floor. If they were all vibating upwards in unison, that would be very ordered. To have them vibrating in all different directions, that is disordered. The ordered states we say have low entropy; the disordered ones, high entropy. And it's always in the nature of the collisions between molecules—because they are so irregular and unsystematic—that they tend to increase the entropy. That way, absolute disorder is the most probable state of any statistical ensemble.'

'You're simply saying that if you leave things to their own devices they tend to get messed up rather than sort themselves out,' suggested Mr Tompkins.

'Yes, you could put it like that,' agreed the professor.

'Not that *Dad* would put it like that. He's just trying to make it sound scientific,' said Maud, sleepily stretching out on her lounger. Arranging a hat over her face to shield her eyes from the Sun, she added in muffled tones, 'But don't let him fool you with his jargon. *Entropy*, I ask you!'

'Thank you, dear,' said the professor indulgently. 'As I was saying: If Maxwell's Demon could be put to work, be would soon put some order into the movement of the molecules the way a good sheep dog rounds up and steers a flock of sheep. Then the entropy would decrease. I should also tell you that according to the so-called Htheorem Ludwig Boltzmann introduced...'

Apparently forgetting he was not talking to a class of advanced students, he continued to ramble on. Using such monstrous terms as 'generalized parameters' and 'quasi-ergodic systems,' he obviously thought he was making the fundamental laws of thermodynamics and their relation to Gibbs' form of statistical mechanics crystal clear. Mr Tompkins was used to his father-in-law lapsing into the habit of talking over his head, so he sipped his coffee and contented himself with trying to look intelligent.

10 Maxwell's Demon

But all this was proving too much for Maud. It was becoming more and more of a struggle to keep her eyes open. She got to thinking of the washing up that still needed doing. So, to throw off her drowsiness she decided to go in and stack the dishes—ready for the men to do the actual washing up later.

'Does madam desire something?' inquired a tall, elegantly dressed butler, bowing as she came into the kitchen.

'No, just go on with your work,' she said, vaguely wondering how they had come to acquire a butler. Presumably her husband had won a fortune on the horses after all, or perhaps he had managed to patent one of his perpetual motion machines. The butler was tall and lean with an olive skin, long, pointed nose, and greenish eyes which seemed to burn with a strange, intense glow. She regarded him as he finished drying the dishes—which she noticed had already been washed. She was curious about the two symmetrical lumps half hidden by the black hair above his forehead. He seemed to bear a striking resemblance to Mephistopheles.

'When exactly did my husband hire you?' she asked, just for something to say.

'Oh, he didn't hire me,' answered the stranger, neatly folding the tea towel. 'As a matter of fact, I came here of my own accord. I do so enjoy making things neat and tidy. I can't *stand* mess. I came to show your distinguished father I am not the myth he believes me to be. I happened to catch sight of the appalling state of the sink as I passed the kitchen door. No offence meant, of course. I am sure someone would have got around to cleaning it up eventually. But I couldn't resist the temptation. I simply *had* to bring a little order to bear. It's my nature—my unnatural nature. Allow me to introduce myself. I am Maxwell's Demon.'

'Oh,' breathed Maud with relief. 'That's all right then; I thought you might be...'

'Yes, I know. I'm often mistaken for him. But have

> 这次，是慕德先睡着并开始做梦了。

> "麦克斯韦妖"，这个物理学家在理论中构建的"妖"，就这样在慕德的梦中登场了。

no fear; I'm quite harmless. The odd practical joke, maybe. But nothing more serious than that. In fact, I was about to play one on your father.'

'What exactly...?' asked Maud, uncertainly. 'I'm not sure my father would appreciate...'

'Oh, don't worry. Just a bit of fun. I simply want to demonstrate that the law of increasing entropy can be broken. And to convince you too, why don't you come along with me?'

Not waiting for her answer, he gripped her elbow. Everything around her suddenly went crazy. All the familiar kitchen objects began to grow with terrific speed— either that, or she and the demon had begun to shrink. She got a last glimpse of the back of a chair covering the whole horizon, before things finally quieted down. She now found herself floating in the air, supported by her companion's strong grip on her arm. Foggy-looking spheres, about the size of tennis balls, joined together in pairs, whizzed by. They came from all directions. Maud was frightened that one of these dangerous looking missiles might hit her.

'What are they?' she asked.

'Air molecules,' replied Maxwell's Demon. 'That one over there is oxygen. And this one...Duck!...' He expertly steered the two of them so as to avoid a collision. 'That one was nitrogen.'

Looking down, Maud caught sight of what looked like a fishing boat. Its deck was completely covered by a quivering heap of glistening fish. Except that when they got closer, it turned out that they were not fish at all, but a seething mass of foggy bails, not unlike those flying past them in the air. The Demon gently but firmly guided her in closer still. Now she could observe how the balls were moving about in a random, pattern less way. Some came floating to the surface, others got sucked down. Occasionally one would come to the surface with such speed it would escape the pull of the others and tear off into space. There were yet other balls flying through the

> 这里讲在慕德眼中一切突然都变得 crazy，一个前提仍是他们与周围事物相对尺度的变化。

> 阅读这段生动形象的描写时，需要有想像力才能在脑海里形成微观世界的奇异图景，不过，要记住这仍只是比拟的"想像"。

10 Maxwell's Demon

IT TURNED OUT THAT THEY WERE NOT FISH AT ALL

air that would dive into the 'soup' and disappear under thousands of other balls.

Looking at the soup more closely, Maud discovered that the balls were of two different kinds. If most looked like tennis balls, there were others, larger and more elongated, that were shaped more like rugby balls. All of them were semi-transparent and seemed to have a complicated internal structure which Maud could not make out.

'Where are we?' gasped Maud. 'Is this Hell!?'

'Of course not,' snapped the Demon, 'I told you before: I'm not who you thought I was. We're simply taking a close look at a very small portion of the liquid surface of the whisky that is about to be drunk by your father—once he gets off the subject of quasi-ergodic systems. The smaller round balls are water molecules; the larger, longer ones are molecules of alcohol. If you care to work out the proportions, you can find out just how strong the drink is that your husband poured.'

> 这就是从微观上看这杯饮料中酒精的浓度了!

Just then Maud spotted what appeared to be a couple of whales playing in the water.

'Atomic whales?' she asked, pointing in their direction.

The Demon looked where she indicated. 'No, no,' he laughed. 'Barley. Very fine fragments of burned barley—the ingredient which gives whisky its particular flavour and colour. Each fragment is made up of millions and millions of complex organic molecules; that's why they are quite large and heavy—compared with individual molecules. In fact, that's very interesting: Note the way they bounce around. See what I mean?'

She nodded. 'Yes. Why are they doing that?'

> 科学家最早找到分子存在的可靠证据,依据的就是这幅图景。

'It's because they are being bumped into by the surrounding molecules. The molecules get their energy from thermal motion. But then they hit the piece of barley. One molecular impact is not going to have much effect. But at any given time there might be more impacts on one side than the other—purely at random—so the impacts add up, and that leads to the barley being pushed in that direction—just for a moment. Then it gets pushed in another direction, and so on. That's why it ends up jiggling about like that.

'In fact, that is how scientists got their first direct proof of the kinetic theory of heat—that matter was made up of molecules moving about. Molecules are too small to be seen down a microscope, but intermediate-sized particles, like that barley fragment, can be seen. And what's more, you can see it doing that jiggle—like dance—*Brownian motion* it's called. So by measuring the extent of its zig—ag path, and applying statistical analysis, physicists were able to get information on the energy of molecular motion—without having to see the individual molecules. Clever, eh?'

The Demon took her right up close to the liquid surface. Now she could see a large transparent block made of numberless molecules fitted neatly and closely together like bricks. Its straight, smooth walls rose up out of the surface of the whisky sea.

'How very impressive!' exclaimed Maud. 'It looks like a glass office block.'

'Not glass. Ice,' corrected the Demon. 'This is part of an ice crystal, one of the cubes in your father's glass. Now if you will sit here for a while,' he said, setting her down on the edge of the ice crystal, so that she perched like an unhappy mountain climber, 'I have work to do.'

Armed with an instrument like a tennis racquet, he dived into the whisky sea. As he swam around, Maud could see him swatting the molecules around him. Darting here and there, he deflected the paths of some molecules one way and others in another direction. Maud could not at first understand the rationale behind his actions. But then his strategy became clear. The fast — moving molecules were being directed to one part of the glass, the slower ones to the opposite side. Maud could not help but admire his speed and dexterity. Such quick thinking! Such skill! Compared with the exhibition she was witnessing,

这就是物理学家想像中的麦克斯韦妖的工作方式。

Wimbledon tennis champions had much to learn.

In a few minutes, the results of the Demon's work became apparent. One half of the liquid surface was now covered by very slowly moving, quiet molecules, while the other became more and more furiously agitated. The number of molecules escaping from the surface in the process of evaporation was increasing rapidly. They were now escaping in groups of thousands together, tearing through the surface as giant bubbles. There were so many of them Maud could get only occasional glimpses of the whizzing racquet or the tail of the demon's dress suit among the masses of maddened molecules.

The whisky-it's boiling

喝酒的人也有被烫着的危险。

Suddenly the demon was at her side.

'Quick!' he said, 'Time to be going before we get scalded.'

With that, he took her elbow once more in his sure grip and propelled her upwards. She now found herself hovering high above the patio, looking down on her father and husband. Her father was springing to his feet.

'Good grief!' he exclaimed, staring bewildered at his whisky glass. 'It's boiling!'

10 Maxwell's Demon

Sure enough, the whisky in the glass was covered with violently bursting bubbles, and a thick cloud of steam rose up in the air.

'Look!' he cried in an awed, trembling voice. 'Here I was telling you about statistical fluctuations in the law of entropy—and now we actually see one! By some incredible chance, possibly for the first time since the Earth began, the faster molecules have all grouped them-selves accidentally on one part of the surface of the water and the water has begun to boil by itself! In the billions of years to come, we will still, probably, be the only people who have ever had the chance to observe this extraordinary phenomenon. What a stroke of luck!'

> 尽管概率极小，这种情形在理论上却是可能的。

As Maud continued to watch from above, she became enveloped in the cloud of steam that had risen from the glass. Soon she could no longer see anything. It became hot and stuffy. She had difficulty breathing. She gasped and struggled.

'Are you all right dear,' Mr Tompkins enquired. He was gently shaking her elbow.'Sounds as though you're suffocating under there.'

She pulled herself together, removing her hat. She blinked at the setting Sun.

'Sorry,' she murmured. 'Must have dropped off.'

She lay there recalling that a friend had recently told her that married people tend to become like each other. She wasn't sure that she relished the idea of having more of the same kind of dreams as her husband. 'Though,' she smiled to herself wistfully, 'We could certainly do with a tame Maxwell's Demon to keep the house in good order.'

> 梦总会醒来。

11 The Merry Tribe of Electrons

快乐的电子部落

A few days later, while finishing his dinner, Mr Tompkins remembered that it was the night of the professor's lecture on the structure of the atom. He had promised to attend, but was particularly tired that evening. The train home from work had been delayed by some breakdown along the line. It had waited for over half an hour outside the station. The weather being hot, the carriage had become unbearably stuffy, and he had arrived home exhausted. He thought he would give the lecture a miss. He hoped his father-in-law might not notice his absence. But just as he settled down to the newspaper to see what was on the television, Maud cut off this avenue of escape by looking at the clock and remarking, gently but firmly, that it was almost time for him to leave.

> 看来爱情也是科普的重要动力。

So, it was he found himself once again on the bench in the university auditorium, together with the usual crowd of students. The professor began...

Ladies and gentlemen:
Last time I promised to give you some details concerning the internal structure of the atom, and how these features account for its physical and chemical properties. You know, of course, that atoms are no longer considered as elementary, indivisible constituent parts of matter. That role has passed now to much smaller particles such as electrons.

11 The Merry Tribe of Electrons

The idea of elementary constituent particles of matter, representing the last possible step in divisibility of material bodies, dates back to the ancient Greek philosopher Democritus. He lived in the fourth century BC. Sitting on some steps one day, he noticed that they were worn. He wondered what was the smallest particle of wearing. Could it be infinitesimally small? In those days it was the custom to try and solve problems by pure thinking. In any case, the question was at that time beyond any possible attack by experimental methods. So it was that Democritus had to search for the correct answer in the depths of his own mind. On the basis of some obscure philosophical considerations, he finally came to the conclusion that it is 'unthinkable' that matter could be divided into smaller and smaller parts without any limit. Thus one must assume the existence of 'the smallest particles which cannot be divided any more'. He called such particles 'atoms'. That is a word which in Greek means 'indivisibles'.

> 古希腊人超越实验的"猜想",是人类智力之力量的证明,是古希腊哲学家天才的杰出表现。

> "原子"一词即由此而来。

I should point out that besides Democritus and his followers, there was undoubtedly another school of Greek philosophy which maintained that the process of divisibility of matter *could* be carried beyond any limit. At this time, and for centuries later, the existence of indivisible portions of matter had to remain a purely philosophical hypothesis.

It was only in the nineteenth century that scientists decided that they had finally found the indivisible building—blocks of matter which had been foretold by the old Greek philosopher more than two thousand years previously. In the year 1808 an English chemist, John Dalton, showed that the relative proportions...

> 到了道尔顿的时代,人们对原子的认识的基础就有所不同了。

From the beginning of the lecture, Mr Tompkins knew his attendance here was a mistake. The urge to rest his eyes, which was ever-present when attending talks, was this evening irresistible. To make matters worse, he had chosen to sit at the end of the row, where he was conveniently able to lean against the lecture theatre wall. Half dozing, half listening, the rest of the lecture became a blur.

> 又进入梦乡了，不过这次还是将人变小，进入微观世界的梦。

With the professor's voice still vaguely echoing in the background, Mr Tompkins experienced the pleasant sensation of floating on air. Opening his eyes he was surprised to find himself dashing through space at what appeared to be a pretty reckless speed. Looking around he saw that he was not alone on this fantastic trip. Near him a number of vague, misty forms were swooping around a large, heavy, nobbly-looking object. These strange beings were travelling in pairs, happily chasing each other along circular and elliptical paths. As they swung round the central object, each of them

They seemed to be dancing a Viennese waltz

11 The Merry Tribe of Electrons

spun like a top; one member of each pair spun one way, and its partner in the opposite direction. For all the world it seemed to Mr Tompkins they were dancing a Viennese waltz. All of which made him feel out of place. Conspicuously, he was the only one of the whole group who had no companion.

这一次，汤普金斯先生干脆变成了"电子"。

'Why didn't I bring Maud along with me?' he wondered gloomily. 'We could have had a wonderful time at this ball.'

The path he was moving along lay outside all the others. He wanted very much to join the rest of the party, but there seemed to be some strange influence that prevented him from getting any closer to them. The uncomfortable feeling of being the odd man out became more pronounced.

Just then, one of the electrons (for by now Mr Tompkins realized he had miraculously joined the electronic community of an atom) was passing close by on its elongated track. He decided to complain about the situation.

'Excuse me, but could you tell me why I don't seem to have a partner, whereas everyone else does?' he shouted across.

'Why? Because this is an odd atom. You are the valency electr-o-o-on...,' called the electron as it turned and plunged back into the dancing crowd.

价电子的概念就这样形象地引入了。

'Valency electrons live alone or find companions in other atoms,' squeaked the high pitched soprano of another electron rushing past him. 'Don't you know anything?'

'If you want a partner fair, jump into chlorine and find one there,' chanted another mockingly.

'I take it you are new here, my son,' said a friendly voice above him. Looking up, Mr Tompkins saw the stout figure of a monk clothed in a brown tunic.

'I am Father Pauli,' went on the monk, moving along the track with Mr Tompkins. 'My mission in life is to keep watch over the morals and social life of electrons in atoms and

原来泡利不相容原理背后还可以有这种神父的背景。

elsewhere. It is my duty to keep these playful electrons properly distributed among the different quantum cells of the beautiful atomic structures erected by our great architect Niels Bohr. To keep order (and to preserve decorum) I never permit more than two electrons to follow the same track. A *ménage à trois* always gives trouble, don't you think? You will note that each electron is neatly paired off with one of the opposite "spin"— a marriage of opposites, if you like. No intruder is permitted if a cell is already occupied by a couple. It's a good rule, and I may add, it has never been broken. The electrons clearly understand that it makes sound sense.'

'Maybe it is a good rule,' objected Mr Tompkins, 'but it is rather inconvenient for me at the moment.'

'I see it is,' smiled the monk. 'But I'm afraid it is just your bad luck, being a valency electron in a sodium atom. It's an odd atom, you see. The electric charge of its nucleus (that big nobbly dark mass over there in the centre)—is enough to hold eleven electrons together. And eleven is an odd number. Half the numbers *are* odd; So it's hardly unusual. I don't see that you can really complain if you turn up late and are the last to attach yourself to an odd atom. You will just have to wait a bit.'

'You mean there is a chance that I can get in later?' asked Mr Tompkins eagerly. 'Kicking one of the old-timers out, for example?'

'Now, now,' admonished the monk, sternly wagging a plump finger at him, 'that is not the way we behave around here. You must learn to be patient. You will find that there is always a chance that some of the inner-circle members will be thrown out by an external disturbance. That way an empty place might become available. However, I wouldn't count on it much, if I were you.'

'They told me I'd be better off if I moved into chlorine,' said Mr Tompkins, discouraged by Father Pauli's words. 'Can you tell me how to do that?'

'Young man, young man!' murmured the monk sorrowfully, 'why are you so insistent on finding company?

11 The Merry Tribe of Electrons

Why can't you appreciate solitude and this Heaven-sent opportunity to contemplate your soul in peace? Why must you electrons always look to the worldly life?' He sighed. 'However, if you insist on companionship, I will help you to get your wish.'

He looked about him intently. After a while he brightened up and began pointing. 'Ah!' he exclaimed. 'Over there. A chlorine atom—and it's approaching us. Look! Even at this distance you can see an unoccupied spot where you would most certainly be welcomed. The empty spot is in the outer group of electrons, the so-called "M-shell", which is supposed to be made up of eight electrons grouped in four pairs. But, as you see, there are four electrons spinning in one direction and only three in the other, with one place vacant. The inner shells (the "K" and "L" shells) are completely filled up. Yes, the atom will be glad to get you and have its outer shell complete too.'

The monk began waving his arm about to attract the chlorine atom's attention, much as one would hail a taxi cab.

神父（泡利）当然不理解汤普金斯想找伴侣的欲望，汤普金斯是凡人，当然也无法欣赏上天赐给他的在平静中关注自己灵魂的机会。

An unoccupied spot in a chlorine atom

'When it gets close, just jump across,' he instructed Mr Tompkins. 'That's what valency electrons usually do. And may peace be with you, my son!' With these words, the priestly father-figure of the electrons suddenly faded into thin air.

Feeling considerably more cheerful, Mr Tompkins gathered his strength for a neck breaking jump into the orbit of the passing chlorine atom. To his surprise he leapt over with an easy grace and found himself in the congenial surroundings of the members of the chlorine M-shell. He was warmly welcomed by the others. Immediately a seductive electron of opposite spin sidlted up to him.

'Delightful of you to join us,' she purred. 'Be my partner, and let's have fun.'

Gliding gracefully along the track together, Mr Tompkins agreed that this really was fun—lots of fun. But one little worry kept stealing into his mind. 'How am I going to explain this to Maud when I see her again?' he thought rather guiltily. But not for long. 'Surely she won't mind,' he decided. 'After all, these are only electrons.'

'Why doesn't that atom you've left go away now?' asked his companion with a pout. 'I hope it doesn't expect to get you back.'

And, as a matter of fact, the sodium atom, with its valency electron gone, *was* sticking closely to the chlorine one.

'Well how do you like that!' said Mr Tompkins angrily, frowning at the atom which had earlier treated him in such an unfriendly, off-hand manner.

'Oh, they're always like that,' said a more experienced member of the M-shell. 'It's not so much the electrons that want you back as the sodium nucleus itself. There's almost always some disagreement between the central nucleus and its electronic escort. The nucleus wants as many electrons around it as it can possibly hold onto with its electric charge, whereas the electrons prefer to have just enough to make the shells complete.

11 The Merry Tribe of Electrons

'There are only a few atomic species—the so-called *rare gases*, or *noble gases* as the German chemists call them—in which the desire of the ruling nucleus and the subordinate electrons are in full harmony. The number of electrons the nucleus can hold on to is just equal to the number needed to give complete shells. Such atoms as helium, neon and argon, for example, are unbelievably smug and self-satisfied. They need neither to expel unwanted extra electrons nor invite new ones to fill up vacancies. So they keep themselves to themselves; they're chemically inert.

> 由此解释了惰性气体性质不活泼的原因。

'But in all other atoms,' this knowledgeable electron continued, 'the electronic communities are always ready to change their membership. In the sodium atom-your former home—the electric charge on the nucleus is enough to hold on to one more electron than is necessary for harmony in the shells. On the other hand, in our atom the normal contingent of electrons is not enough for complete harmony. And that's why we welcome your arrival—in spite of the fact that your presence here overloads our nucleus. As long as you stay here, our atom is no longer electrically neutral; it has your extra negative electric charge. The sodium atom you left is now short of an electron, so overall it now has a positive electric charge. That's why it continues to stand by; it's held by the force of electric attraction between its positive charge and our negative one. I once heard our great priest, Father Pauli, say that such atomic communities, with extra electrons or electrons missing, are called negative and positive "ions". He also uses the word "molecule" for groups of two or more atoms bound together by these electric forces. This particular combination of sodium and chlorine atoms he calls a molecule of "table salt"—whatever that might be.'

> 也由此解释了我们每天都吃的食盐，也即氯化钠分子生成的化学原因。

'Do you mean to tell me you don't know what table salt is?' said Mr Tompkins, forgetting to whom he was talking. 'Why, that's what you put on your scrambled eggs at breakfast.'

> 身为被吃食品的组成部分,当然不知道吃它的人看它是什么样子了。

'What are "scram bulldeggs", and what is "breakfust"?' asked the electron.

Mr Tompkins was at a loss to know what to say. He realized the futility of trying to explain to his companions even the simplest details of the lives of human beings. Fortunately the informative electron was not really interested in anything he had to say about human life, being too busy showing off her own knowledge of the world of electrons.

'You must not think,' she continued, 'that the binding of atoms into molecules is always accomplished by one valency electron alone. There are atoms, like oxygen for example, which need two more electrons to complete their shells, not just the one that chlorine needs. And there are also atoms which need three electrons, and even more. On the other hand, in some atoms the nucleus holds two or more extra—or valency—electrons. When such atoms meet, there is quite a lot of jumping over and binding to do. As a result, you get quite complex molecules, often consisting of thousands of atoms. There are also the so-called "homopolar" molecules, that is, molecules made up of two identical atoms, but that is a very unpleasant situation.'

'Why unpleasant?' asked Mr Tompkins.

> 这本书要是在今天由中国科普作家来写,也许对这种电子的比喻就不是乒乓球,而是在舞台上跳来跳去的"超级女生"了。

'Too much work,' replied the electron. 'It's too much work to keep them together. Some time ago I got the job. I didn't have a moment to myself. It's nothing like the way it is here where the valency electron just enjoys herself while the deserted atom stands by. No sir! To keep two identical atoms together, you have to jump to and fro, from one to the other and back again. Back and forth, back and forth, all the time. It's as bad as being a ping pong ball.'

Mr Tompkins was rather surprised to hear the electron, which did not know what scrambled eggs were, speak so knowledgeably of ping pong, but he let it pass.

'I'll never take on that job again!' declared the electron. 'I'm quite comfortable where...' Her voice

11   The Merry Tribe of Electrons

trailed off as something caught her attention. 'Hey! Did you see that?! Hah! An even better place. So lo-o-o-ong!'

A startled Mr Tompkins watched as the electron, with a giant leap, swooped towards the interior of the atom. It seems that one of the electrons of the inner circle had been thrown clear of the atom by some foreign high-speed electron which had unexpectedly penetrated into their system. A cosy place in the 'K' shell was now wide open. Chiding himself for missing this opportunity to join the inner circle, Mr Tompkins now watched with great interest the course of the electron he had just been talking to. Deeper and deeper into the atomic interior this happy electron sped, bright rays of light accompanying her triumphant flight. Only when it finally reached the internal orbit did this almost unbearable radiation cease.

电子跃迁，带来辐射。

'What was that?' asked Mr Tompkins, his eyes aching. 'That flash of light. What was that all about?'

'Oh that's just the X-ray emission,' explained his orbit companion. 'You get it when there's a transition. Whenever one of us succeeds in getting deeper into the interior of the atom, the surplus energy must be emitted in the form of radiation. That lucky girl made quite a big jump. That let loose a lot of energy. More often we have to be satisfied with smaller jumps, here in the atomic suburbs, and then our radiation is called "visible light"—at least that's what Father Pauli calls it.'

'But this X-ray light, it was also visible,' protested Mr Tompkins. 'I *saw* it. Why isn't that called "visible" too?'

'We're electrons. We're susceptible to *any* kind of radiation. But Father Pauli tells us that there exist gigantic creatures, "Human Beings", he calls them, who can see light only when it falls within a narrow wave length interval. He told us once that it wasn't until some human called Roentgen came along that they even knew that there was such a thing as X-rays. These humans don't sound very clever to me. Anyway, having at long last discovered them, I understand they use them quite a lot

"电眼"看人！

now—in something called "medicine".'

'Oh yes. I know quite a lot about that,' said Mr Tompkins. 'Medicine is where we try to... I mean, where *humans* try to help those...'

The electron yawned rudely. 'I shouldn't bother. I really don't care. Come on, let's dance.' With that she took his hand and they whirled along their trajectory.

For quite some time Mr Tompkins was content to enjoy the pleasant sensation of diving through space with the other electrons in a kind of glorified trapeze act. Then, all of a sudden, he felt his hair stand on end, an experience he had felt once before during a thunder storm in the mountains. It was clear that a strong electrical disturbance was approaching their atom, breaking the harmony of the electronic motion, and forcing the electrons to deviate seriously from their normal tracks. He was later to learn that it was only a wave of ultraviolet light passing through the spot where this particular atom happened to be, but to the tiny electrons it was a terrific electric storm.

'Hold on tight!' yelled someone, 'or you'll be thrown out by photoelectric-effect forces!' But it was already too late. Mr Tompkins was snatched away from his companions and hurled into space at a terrifying speed. It was as though he had been neatly extracted from the atom by tweezers. Breathlessly he hurtled further and further through space, tearing past all kinds of different atoms so fast he could hardly distinguish the separate electrons. Suddenly a large atom loomed up right in front of him and he knew that a collision was unavoidable.

'Pardon me, but I am photoelectric-effected and cannot...,' began Mr Tompkins politely, but the rest of the sentence was lost in an ear-splitting crash as he ran head on into one of the outer electrons. The two of them tumbled head over heels. Mr Tompkins lost most of his speed in the collision, and now found himself trapped in new surroundings.

电子眼中的"光电效应"。

11 The Merry Tribe of Electrons

Having regained his breath, he examined his environment. He found himself to be hemmed in on all sides by atoms. They were much larger than any he had seen before. He could count as many as twenty-nine electrons in each of them. If he had known his physics better he would have recognized them as atoms of copper, but at these close quarters the group as a whole did not look like copper at all. He noted that they were not only spaced rather close to one another, but they were arranged in a regular pattern which extended as far as he could see.

> 在电子眼中金属内部的景象。

But what surprised Mr Tompkins most was the fact that these atoms did not seem to be very particular about holding on to their quota of electrons, particularly their outer electrons. In fact the outer orbits were mostly empty, and crowds of unattached electrons were drifting lazily about, stopping from time to time, but never for very long, on the outskirts of one atom or another. They reminded him of gangs of youths hanging around street corners and wandering aimlessly down the street in the evening with nothing to do.

> 幸福的自由电子!

Tired after his breakneck flight through space, Mr Tompkins tried at first to get a little rest on a steady orbit of one of the copper atoms. However, he was soon infected with the prevailing vagabondish feeling of the crowd, and he joined the rest of the electrons in their nowhere-in-particular motion.

'I must say things don't seem to be very well organized here,' he commented to himself. 'There are too many electrons not attending to their business– leading aimless lives. I wonder if Father Pauli knows about this lot.'

'Of course I do,' said the familiar voice of the monk. He had suddenly materialized from nowhere. 'There's no problem. These electrons are not disobeying any rules. In fact, they're doing a very useful job. If all atoms cared as much about holding on to their electrons as some of them do, there would be no such thing as electric conductivity. No electric appliances, no electric lights, no computers, TVs, radios.'

'Are you saying these electrons—these *wandering* electrons—are responsible for electric current?' asked Mr Tompkins. 'I don't see how. It's not as though they are moving in any particular direction.'

'You wait and see,' said the monk. 'All it requires is for someone to press the switch. And by the way, I don't know why you are using the word "they"; it ought to be "we". You seem to forget that you are a conducting electron yourself.'

'As a matter of fact I'm getting quite tired of being an electron,' said Mr Tompkins. 'It was fun to begin with, but the novelty is fast wearing off. I've come to the conclusion I am not cut out to be following these rules, and being knocked around for ever.'

'Not necessarily forever,' countered Father Pauli somewhat testily. He clearly did not expect 'lip' from a mere electron. 'There is always the chance that you will get annihilated.'

> 莎士比亚名言的"电子版"：存在，还是湮灭，这是个问题。

'Annihilated!?' exclaimed Mr Tompkins in alarm. 'But I thought electrons were eternal.'

'That is what physicists used to believe,' agreed Father Pauli, 'but now they know better. Electrons can be born, and die, just like humans. Not that they die of old age, of course. Death comes *suddenly*, without warning, through collisions.' He smiled as he relished the disconcerting effect his words were having on Mr Tompkins.

'I had a collision only a short while ago. It was a pretty bad one too,' said Mr Tompkins recovering a little confidence. 'But it didn't put me out of action. Don't you think you're being just a wee bit over-dramatic?'

'It's not a question of how *forcibly* you collide,' Father Pauli corrected him. 'It's all to do with *what* you collide with. If I am not mistaken, your recent collision was with another negative electron—one very similar to yourself. There's not the slightest danger in such an

11 The Merry Tribe of Electrons

encounter. In fact, you could butt each other like a couple of rams for years and no harm could be done. But there is another breed of electron: the positively charged ones. Those are the ones to watch out for. The positive electrons, or *positrons*, look exactly the way you do. When you see one approaching, you think it's just another innocent member of your tribe. So, you go ahead and greet him. But then you find that, instead of your negative charges pushing you away slightly to avoid too close a collision, his positive charge attracts your negative one, and he pulls you right in. And then it is too late to do anything.'

用中国的老话讲，那就是撞上"冤家"了。

'Why? What happens then?' asked Mr Tompkins.

'You get eaten up. Destroyed.'

'Oh! And how many poor ordinary electrons can one positron eat up?'

'Fortunately only one. In destroying a negative electron, the positron also destroys itself. I suppose you could say they have a kind of death wish—always on the lookout for a partner with whom they can enter into a suicide pact. Positrons do not harm each other; but as soon as a negative electron comes their way, it hasn't much chance of surviving.'

'Then it's lucky I haven't run into one of these monsters yet,' said Mr Tompkins nervously. 'I hope they're not very numerous. Are they?'

'No, no. They don't hang around long enough for that; always looking for trouble and so vanish very soon after they're born. If you wait a minute, I shall probably be able to show you one.'

Father Pauli looked about him for a few minutes, then exclaimed, 'Yes, there we are!' He pointed to a distant heavy nucleus. 'Can you see? That's a positron being born.'

The atom at which the monk was pointing was evidently undergoing a strong electromagnetic disturbance owing to some vigorous radiation falling on it from outside. It was a much more violent disturbance than the one which threw Mr Tompkins out of his chlorine atom. The atom's family of electrons was being blown away like dry leaves in a hurricane.

'Look closely at the nucleus,' said Father Pauli. Concentrating his attention, Mr Tompkins saw something very unusual. Close to the nucleus, inside the inner electronic shell, two vague shadows were quickly taking shape. A moment later, Mr Tompkins saw two glittering brand new electrons rushing at great speed away from their birthplace.

'But I see two of them,' said Mr Tompkins, excitedly.

'That's right,' agreed Father Pauli. 'Electrons are always born in pairs. They're electrically charged, so you have to produce two at the same time—one with positive charge, the other with negative—otherwise it would contradict the law of conservation of electric charge. So, the action of that strong gamma ray on the nucleus has produced an ordinary negative electron as well as the positron.'

'Oh, that's not so bad then,' commented Mr Tompkins. 'If the birth of each positron is accompanied by an extra negative electron, then that means when the positron destroys a negative one later, we're back to where we were—as far as the total number of electrons is concerned. So it doesn't lead to the extinction of the electronic tribe. and I...'

'If I were you, I would watch out for that positron,' interrupted the monk.

'Which one's the positron?' asked Mr Tompkins. 'They look the same to me.'

'Not sure. But one of them's coming our way.'

He brusquely shoved Mr Tompkins aside while the newborn particle whistled by. No sooner had it passed, than it crashed into another electron. There were two blinding flashes of light—then nothing!

'Well, I guess that answers your question,' smiled the monk.

Mr Tompkins's relief at escaping the clutches of the

11 The Merry Tribe of Electrons

murderous positron, however, were short lived. Before he had time to thank Father Pauli for his quick thinking, he abruptly felt himself being pulled. He and all the other wandering electrons had been galvanized into action and were all being propelled in the same direction.

'Hey! What's happening now?' he cried.

'Someone must have pressed the light switch. You're on your way to the filament in the light bulb,' called the monk, who by now was fast disappearing into the distance, 'Bye! Nice talking to you.'

At first the journey was quite pleasant and effortless—like being transported on a moving walkway at an airport. Mr Tompkins and the other loose electrons were gently weaving their way through the lattice of atoms. He tried to get into conversation with a nearby electron.

'Quite relaxing this, isn't it,' he remarked.

The electron shot him a menacing look. 'Huh! You're obviously new to this circuit. Just you wait till we get to the rapids.'

Mr Tompkins did not know what this meant, but did not like the sound of it. He did not have long to wait to find out. Suddenly the channel through which they were passing narrowed. The electrons were now crushed together as they squeezed along. It became hotter and hotter, and brighter and brighter.

> 也许这就是电子在电流中受的"电刑"吧。

幸好慕德没看见!

'Brace yourself!' muttered his companion as she came crashing in from the side.

Mr Tompkins awoke to find that the woman sitting next to him on the lecture theatre bench had also dozed off, and had slumped sideways onto him, pushing him up against the wall.

11½ The Remainder of the Previous Lecture through which Mr Tompkins Dozed
上一次演讲中汤普金斯先生因为睡着而没有听到的那部分

... In fact, in the year 1808, an English chemist John Dalton showed that the relative proportion of the chemical elements needed to form a complicated chemical compound is always a ratio of whole numbers. He interpreted this rule as indicating that all chemical compounds are built up from particles representing simple chemical elements. The failure of medieval alchemy to turn one chemical element into another supplied supporting evidence of the apparent indivisibility of these particles. So, without much hesitation they were given the old Greek name: 'atoms'. Although we know now that these 'Dalton's atoms' are not at all indivisible (they are, in fact, formed from still smaller particles), the name 'atom' stuck.

> 这篇演讲可以说是对前一篇汤普金斯先生打瞌睡时所梦到的原子中生动图景的一种比较学术化的说明。

Thus the entities called 'atoms' by modern physics are not at all the elementary and indivisible constituent units of matter imagined by Democritus, and the term 'atom' would actually be more correct if it were applied to such smaller particles as electrons and quarks from which 'Dalton's atoms' are built. (Quarks, incidentally, are the ultimate constituents of atomic nuclei; I shall be having more to say about them at a later date.) Such a change of names at this late stage would cause much confusion. For this reason we retain the old name of 'atoms' in Dalton's sense, and refer to particles such as electrons and quarks as 'elementary particles'. This latter name indicates, of course, that we believe at present that these smaller particles are *really* elementary and indivisible in Democritus' sense of the word. You may well ask whether history will not repeat itself, and whether in the further progress of science, the elementary particles of modern physics will prove to be quite complex. My answer is that, although there is no absolute guarantee that this will not happen, there are very good reasons to believe that this time we are right.

> 在这里，可以看出作者伽莫夫并不相信物质的"无限可分"。

There are ninety-two different kinds of atoms (corresponding to ninety-two different chemical elements), and each kind of atom possesses rather complicated characteristic properties. This in itself invites the suggestion that they might have rather complicated structures constructed out of more elementary ones.

How are Dalton's atoms to be built up from the elementary particles? The first step towards answering this question was taken in 1911 by the celebrated British physicist Ernest Rutherford (later Lord Rutherford of Nelson). He was studying the structure of atoms by bombarding them with alpha particles. (You recall these are the nuclei of helium atoms.) These positively charged particles are emitted in the process of disintegration of radioactive elements. Rutherford observed the deflection (that's to say, the scattering) of these projectiles after their passage through a piece of matter. He found that whereas most of the projectiles were able to pass through with very little deviation, a few recoiled through exceptionally large angles. It was as though they had scored a bullseye on something very small and highly concentrated within the atom. In this way, he came to the conclusion that all atoms must possess a very dense, positively charged central core, or *nucleus*. This he envisaged as being surrounded by a rather rarefied cloud of negative electric charge.

> 这是卢瑟福所做的在人类研究原子结构的历史中非常经典的实验。

11½ The Remainder of the Previous Lecture through which Mr Tompkins Dozed

It was later discovered that the atomic nucleus is made up of a certain number of positively charged *protons* and electrically neutral *neutrons*. These are so similar to each other (apart from their charge) that they are known under the collective name: *nucleons*. They are held tightly together by a short-ranged, powerful cohesive force known as the *strong nuclear force*. It gets its name because it is strong enough to keep protons bound within the nucleus despite the repulsive force acting between their positive charges.

As for the surrounding cloud, this consists of negative electrons swarming around under the restraining influence of the electrostatic attraction exerted by the positive charge of the protons in the nucleus. (You recall, of course, that like charges repel, whereas unlike charges attract.) The number of electrons forming the atomic cloud varies from one type of atom to another, and determines all the physical and chemical properties of a given type of atom. The number of electrons varies along the natural sequence of chemical elements from one （for hydrogen）up to ninety-two （for the heaviest naturally occurring element: uranium）.

In spite of the apparent simplicity of Rutherford's atomic model, its detailed understanding turned out to be anything but simple. For example, what was to stop all the electrons being quickly drawn into the nucleus by the electrostatic attraction? According to classical ideas, the only explanation must be that the electrons are avoiding the nucleus in much the same way as the planets in the Solar System avoid being pulled into the Sun. This they do by moving in orbits about the centre of attraction(in that case, gravitational attraction). But unfortunately, classical physics also says that when the orbitting body is electrically charged, it will progressively radiate energy away—a form of light-emission. It was calculated that, due to these steady energy losses, all the electrons forming an atomic cloud should collapse on the nucleus

> 这正表现出古典物理理论的问题，即无法与新发现的实验结果（也即卢瑟福原子模型）相一致。

within a negligible fraction of a second. This seemingly sound conclusion of classical theory stands, however, in sharp contradiction to the empirical fact that atomic clouds are, on the contrary, quite stable. Instead of collapsing on the nucleus, atomic electrons continue their motion around the central body for an indefinite period of time. Thus we see that a deep-rooted conflict arises between the basic ideas of classical mechanics, and the empirical data concerning the mechanical behaviour of atoms.

It was this contradiction that brought the famous Danish physicist Niels Bohr to the realization that classical mechanics, which claimed for centuries a privileged and secure position in the system of natural sciences, should from now on be considered as a restricted theory. It is applicable only to the macroscopic world of our everyday experience, but fails badly in its application to the much more delicate types of motion taking place within atoms.

As the tentative foundation for a new type of mechanics (one that was eventually to flower into the quantum mechanics I discussed in an earlier lecture), Bohr proposed that *from the infinite variety of orbits theoretically possible in classical theory, only a few specially selected ones are available to electrons orbiting an atomic nucleus.* These permitted orbits, or trajectories, are selected according to certain mathematical conditions, known as the *quantum conditions* of the Bohr theory.

I shall not enter here into a detailed discussion of these quantum conditions, but will mention only that they have been chosen in sush a way that all the restrictions imposed by them become of no practical importance in cases where the mass of the moving particle is much larger than the masses we encounter in atomic structures. Thus, when the new mechanics is applied to macroscopic objects, such as orbiting planets, one gets the same results as the old classical theory. This so-called *principle*

11½ The Remainder of the Previous Lecture through which Mr Tompkins Dozed

of correspondence ensures, for example, that although a planet has only certain orbits about the Sun open to it, these are so numerous and so close to each other that the restriction is not apparent. It therefore becomes easy to form the impression that there is no restriction on the type of orbit permitted. It is only in the case of tiny atomic mechanisms that the difference between adjacent permitted states becomes so marked that one cannot any longer ignore the fact that restrictions on the trajectories do indeed apply, and the disagreement between the two theories becomes marked.

Without going into any details, let me indicate the kind of results that stem from Bohr's theory. On this slide, I have shown (on a greatly magnified scale, of course), the system of circular and elliptical orbits, representing the only types of motion permitted by Bohr's quantum conditions for the electrons of this particular atom. Classical mechanics would allow the electron to move at *any* distance from the nucleus and puts *no restriction* on the eccentricity (i. e. elongation) of its orbit. In contrast, the selected orbits of Bohr's theory form a discrete set with all their characteristic dimensions sharply defined. The combination of a number and a letter next to each orbit indicates the name of that orbit, using the type of classification scheme that has come to be adopted. You may notice, for example, that the larger numbers correspond to the orbits of larger diameter.

Although Bohr's theory of atomic structure turned out to be extremely fruitful in the explanation of various properties of atoms and molecules, the fundamental notion of discrete quantum orbits remained obscure. The deeper one tried to go into the analysis of this unusual restriction on the classical theory, the more confused the overall picture became. It became clear that the fundamental problem with Bohr's theory was that it was based on restricting the results of classical physics by a system of additional conditions which were in principle

也就是说，此时物理学正需要一场彻底的革命！

图中所画的玻尔理论描述的（也是在卢瑟福的实验基础上建立的）原子结构模型，现如今已被广泛地用作科学的象征。它形象，但却并不够先进。

11½ The Remainder of the Previous Lecture through which Mr Tompkins Dozed

quite foreign to the whole structure of classical theory. What was required was a complete rethinking of the underlying physics.

The correct solution came thirteen years later, in the form of socalled *quantum mechanics* (alternatively known as *wave mechanics*). This modified the entire basis of classical mechanics. In spite of the fact that the system of quantum mechanics might at first seem still crazier than Bohr's old theory, this new micro-mechanics represents one of the most consistent and accepted parts of the theoretical physics of today. I have already, in a previous lecture, talked at length about the new mechanics-in particular the notions of 'indeterminacy' and 'spreading out trajectories'. So, I will not repeat myself here. Instead, let us look a little more closely at how those ideas apply to the problem of atomic structure.

In this second slide, you see the way in which the motion of atomic electrons is visualized by quantum mechanics from the point of view of 'spreading out

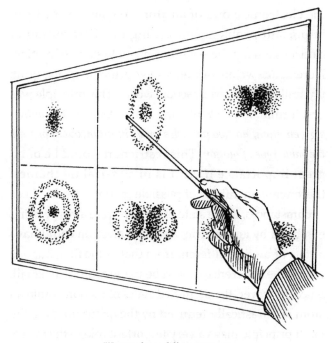

这一幅插图中的原子形象,更接近于"真实",但却又不那么"艺术"了。

We now have diffuse patterns

trajectories'. This picture depicts the types of motion corresponding to those represented classically in the previous diagram (apart from the fact that, for greater clarity, I am now showing each type of motion drawn separately). You can see how, instead of the sharp-lined trajectories of Bohr's theory, we now have diffuse patterns consistent with the fundamental *uncertainty principle*. The notation of the different states of motion is the same as on the previous diagram. In fact, if you compare the two (and stretch your imagination slightly), you will notice that our cloudy forms reflect, to some extent, the general features of the old Bohr orbits. For instance, larger numbers correspond to larger patterns, circular orbits with spherical shapes, elliptical orbits with elongated patterns. These diagrams show what happens to the good old fashioned trajectories of classical mechanics when the quantum is at play. Although it takes a little getting used to, scientists working in the microcosmos of atoms have no difficulty accepting this picture.

So much for the possible states of motion in the electronic cloud of an atom. We now come to an important problem concerning the distribution of electrons among these various possible states of motion. Here again we encounter a new principle—one quite unfamiliar in the macroscopic world. This principle was first formulated by Wolfgang Pauli. It states that *within a given atom, no two particles may simultaneously possess the same type of motion*. This restriction would be of no great importance if, as it is in classical mechanics, there were an infinity of possible motions. Under those circumstances, if one state of motion were already being exhibited by an electron, a second electron could have a state of motion different from that of the first, but one where the differences could be made arbitrarily small. Since, however, the permitted states of motion within an atom are drastically reduced by the quantum laws, the Pauli principle plays a very important role in the micro-

> 这里说的电子"云"的稀疏，实际上是对电子出现在该处的概率大小的形象表示。

11½ The Remainder of the Previous Lecture through which Mr Tompkins Dozed

world. It means, for example, that if states of motion close to the nucleus are already filled by electrons, additional electrons have to occupy states that lie significantly further out from the nucleus. This prevents them from crowding together in one particular spot.

You must not conclude, however, from what I have said so far, that each of the diffuse quantum states of motion represented on my diagram may be occupied by one electron only. In fact, quite apart from the motion along its orbit, each electron is also spinning around its own axis—in much the same way as the Earth spins about its North-South axis in addition to orbiting the Sun. It would not distress Dr Pauli at all if two electrons move along the same orbit, provided they spin in different directions. Now the detailed study of electron spin indicates that the speed of their rotation around their own axis is always the same, and that the direction of this axis must always be perpendicular to the plane of the orbit. This leaves only two different possibilities of spinning; these can be characterized as 'clock-wise' and 'anti-clockwise'.

> 这里讲的电子的旋转（自旋，即 spin），也是一种形象化的描述。

Thus the Pauli principle as applied to the quantum states in an atom can be reformulated in the following way: *each quantum state of motion can be occupied by not more than two electrons, in which case, the spins of these two particles must be in opposite directions.* Accordingly, as we proceed along the natural sequence of elements towards the atoms with larger and larger numbers of electrons, we find different quantum states of motion being gradually filled with electrons, starting from those lying closet to the nucleus, to those lying further out.

It must also be mentioned in this connection that, from the point of view of the strength of their binding, different quantum states of atomic electrons can be united in separate groups (or shells) of states with approximately equal binding. When we proceed along the natural sequence of elements, one group after another is filled,

> 在此篇报告中，虽然学术一些，但仍用了不少比喻性的描述。其实严格地讲，用日常观念来理解这些比喻性的描述，都是

> 不够严格的。可是，又有什么办法呢?

and, as a consequence of their subsequent filling of electronic shells, the properties of the atoms also change periodically. This is the explanation of the well-known periodic properties of elements, discovered empirically by the Russian chemist Dimitrij Mendeleéff.

12 Inside the Nucleus

原子核内容

The next lecture which Mr Tompkins attended was devoted to the study of atomic nuclei. The professor began:

Ladies and gentlemen：

这一次的演讲确实是开门见山了，没有插曲，也没有梦，而是直奔主题。

Digging deeper and deeper into the structure of matter, we shall now try to penetrate with our mental eye into the interior of the atomic nucleus—that mysterious region occupying only one thousand million millionth part of the total volume of the atom itself. Yet, in spite of the almost incredibly small dimensions of our new field of inves-tigation, we shall find it packed with fascinating activity.

同前面讲原子结构的演讲相比，这次演讲的内容更深入一层，深入到原子核的层次。

Entering the nuclear region from the thinly populated electron cloud of the atom, we are at once surprised by the extremely ove-crowded state of the local population. The nucleus, in spite of its relatively small size, contains about 99.97% of total atomic mass. Here the particles rub shouldeis with one another— or they would if they had shoulders. In this respect the picture presented by the nuclear interior is somewhat similar to that of a liquid, such as water—except that in place of water molecules, we here encounter much smaller particles: the protons and neutrons. As far as their geometrical dimensions are concerned, nucleons possess a diameter of about 0.000,000,000,000,001 metre.

The nucleons are packed close together due to the action of the strong nuclear force. It functions in rather the same way as the forces acting between the molecules in a liquid. And, just as in liquids, those forces, while preventing the particles from being completely separated,

> 物理学家在研究原子核时，确实曾用过"液滴模型"。

do not hinder their displacement relative to one another. Thus nuclear matter possesses a certain degree of fluidity. When undisturbed by any external forces, it assumes the shape of a spherical drop, just like an ordinary drop of water.

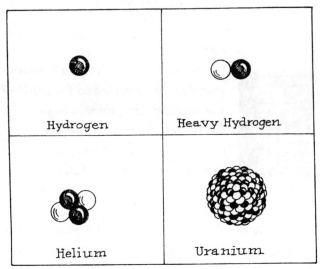

Nuclei

In the schematic diagram which I am going to draw for you now, you see different types of nuclei built from protons and neutrons. The simplest is the nucleus of hydrogen, which consists of just one proton. In contrast, the most complicated uranium nucleus consists of 92 protons and 142 neutrons. Of course, you must consider these pictures only as a highly schematic presentation of the actual situation, since, owing to the fundamental uncertainty principle of the quantum theory, the position of each nucleon is actually 'spread out' over the entire nuclear region.

As I have said, particles forming an atomic nucleus are held together by strong cohesive forces. But in addition to these attractive forces, there are also forces of another kind acting in the opposite direction. As you know, the protons, which form about one half of the total nuclear population, all carry positive electric charge. They are thus mutually repelled from one another by the Coulomb electrostatic forces. For the light nuclei, where the electric charge is comparatively small, this Coulomb repulsion is of little consequence. But in the case of heavier, highly charged nuclei, Coulomb forces offer serious competition to the attractive strong nuclear force. The latter is short-ranged and therefore operates only between neighbouring nucleons. The electrostatic force, on the other hand, is long-ranged. That means a proton on the periphery of the nucleus will be attracted only by its immediate neighbours, but repelled by *all* the other protons in the nucleus. The addition of more protons would progressively build up the repulsive force, without a compensating increase in the strong attractive force (there being a physical limit to how many 'shoulders' a proton can rub at the same time). Above a certain size, the nucleus is no longer stable, and is apt to eject some of its constituent parts. And that is exactly what happens to a number of elements located at the very end of the periodic system of classifying the elements developed by Mendeleéff: those known as the 'radioactive elements'.

> 与家庭大就倾向于更容易分家一样。

From all this, you might conclude that these heavy unstable nuclei should emit protons (since neutrons do not carry any electric charge, and are therefore not subject to the Coulomb repulsive forces). Experiments show, however, that the particles actually emitted are alpha particles. The reason for this specific grouping of nuclear constituent parts is that this particular combination of two protons and two neutrons is especially stable; it is very efficiently locked together. It is therefore easier to throw out the whole group at once, rather than break it into separate protons and neutrons.

The phenomenon of radioactive decay was first discovered by the French physicist Henri Becquerel. Its interpretation as the result of spontaneous disintegration of atomic nuclei was given by the British physicist Lord Rutherford—a name I have already mentioned in other connections. Science owes a great debt to Rutherford for

> 放射性的发现,是19世纪末物理学中"三大发现"之一。另外两大发现分别是"电子"和"X射线"。

his important discoveries in the physics of the atomic nucleus.

One of the peculiar features of the process of alpha decay consists of the sometimes extremely long periods of time needed for alpha particles to make their 'getaway' from the nucleus. For *uranium* and *thorium* this period is measured in billions of years; for *radium* it is about sixteen centuries. There are some elements for which decay takes place in a fraction of a second, but even in these cases, the life-span is still very long compared with the rapidity of intra-nuclear motion. So, we have to ask what constrains alpha particles to stay inside the nucleus, sometimes for many billions of years, when the repulsive forces are clearly strong enough to boot them out. And having already stayed so long, what finally triggers their expulsion?

> 这里又一次谈到了前面曾屡次提到的卢瑟福的著名实验。

To answer this, we must first learn a little more about the comparative strength of the cohesive nuclear force, and the repulsive electrostatic force. A careful experimental study of these forces was made by Rutherford. He used the so-called 'atomic bombardment' method. In his famous experiments at the Cavendish Laboratory, Rutherford directed a beam of fast moving alpha particles, emitted by some radioactive substance, and observed the deviations (the scattering) of these atomic projectiles resulting from their collisions with the nuclei of the bombarded substance. These experiments confirmed that, while at great distances from the nucleus the projectiles are repelled by the long-range electric force of the nucleus, this repulsion changes into a strong attraction if the projectile manages to come very close to the outer limits of the nuclear region. You can say that the nucleus is somewhat analogous to a fortress surrounded on all sides by a high, steep bulwark, preventing the particles from getting in as well as from getting out.

But the most striking result of Rutherford's experiments was the discovery that the alpha particles getting out of the nucleus in the process of radioactive decay, as well as the projectiles penetrating the nucleus from outside, actually possess *less energy* than would correspond to the top of the bulwark, or the *potential barrier*, as we usually call it. This stood in complete contradiction to all the fundamental ideas of classical mechanics. Indeed, how can you expect a ball to roll over a hill if you have thrown it with less energy than is necessary to get to the top of the hill? Classical physics could only suppose that there must have been some mistake in Rntherford's experiments.

But there was no mistake. The situation was clarified simultaneously by George Gamow and by Ronald Gurney and E.U.Condon. They pointed out that there was no difficulty, provided one took into account quantum theory. As we have noted, we know that quantum physics rejects the well-defined linear trajectories of classical theory, and replaces them with diffuse ghostly trails. And, just as a good old-fashioned ghost could pass without difficulty through the thick masonry walls of an old castle, these ghostly trajectories can penetrate through potential barriers which seem to be quite impenetrable from the classical point of view.

可是在我们日常的观念中，确定的轨道似乎是天经地义的。这也说明了量子力学的革命性。

And do not for one moment think I am joking. The penetrability of potential barriers for particles with insufficient energy comes as a direct mathematical consequence of the fundamental equations of the new quantum mechanics. It represents one of the most important differences between the new and old ideas about motion. But, although the new mechanics permits such unusual effects, it does so only with strong restrictions: in most cases the chances of crossing the barrier are extremely small, and the imprisoned particle must throw itself against the wall an almost incredible number of times before its attempt finally succeeds. The quantum theory gives us exact rules concerning the

在量子力学中，这种像"崂山道士"穿墙而过的事情，在微观世界中不再是不可能的。

calculation of the probability of such an escape; it has been shown that the observed periods of alpha decay are in complete agreement with the expectation of the theory. Also, in the case of projectiles which are shot into the nucleus from the outside, the results of quantum-mechanical calculations are in very close agreement with the experiment.

Before going any further, I want to show you some photographs representing the process of disintegration of various nuclei when hit by high-energy atomic projectiles. The first is an old cloud chamber picture. I should explain that with these subatomic particles being so small, one is not able to see them directly, even under the most powerful microscope. So you must not expect me to provide you with actual photographs of them. No, we have to be cunning.

Consider a vapour trail left by a high-flying aircraft. The plane itself might be so high it is difficult to see; indeed it might no longer be there at all. But we know about it from the vapour trail it has left behind. C. R. T. Wilson saw in this a simple way of rendering subatomic nuclei 'visible'. He built a chamber containing gas and vapour. Using a piston, he suddenly expanded the gas. This caused an immediate drop in temperature, plunging the vapour into a super-saturated condition; the vapour was all set to form a cloud. But clouds can't just start forming. They have to have some centres on which to condense (otherwise, why would a drop begin to form in one location rather than another?). What normally happens in cloud formation is that dust particles present in the atmosphere become the preferential centres upon which condensation can begin. The clever thing about the Wilson cloud chamber, however, is that he excluded any dust. So where were the droplets to form? It so turns out that when a charged particle moves through a medium, it ionises atoms in its path (that is to say, it ejects electrons from their atoms). These ionised atoms make good centres

12 Inside the Nucleus

from which droplets can grow. So what happened in the chamber was that wherever the charged particles went (leaving their trail of ionised atoms behind them) there formed a trail of droplets which in a fraction of a second had grown to visible size and could be photographed. And that is what is happening in this slide. From the left you see numerous trails of droplets. Each trail is caused by an alpha particle radiating from a powerful alpha-ray source (which is not shown in the picture). Most of these particles are passing through the field of vision without a single serious collision, but one of them—just below the

> 威尔逊云室的巧妙构思让我们"看"到了粒子的运动。不过，实际上，我们"看"到的，只是根据理论分析推论出的粒子运动后留下的"痕迹"。

There formed a trail of droplets

middle of the picture—has succeeded in hitting a nitrogen nucleus. The track of the alpha particle stops at the collision point, and from the same place you can see two other tracks emerging. The long thin track going up the picture belongs to a proton kicked out from the nitrogen nucleus. The short heavy one represents the recoil of the nucleus itself. This isn't, however, a nitrogen nucleus any more, since by losing a proton and absorbing the incident alpha particle, it has been transformed into a nucleus of oxygen. Thus we have here an alchemic transformation of nitrogen into oxygen with hydrogen as a by-product. I am showing you this because it is the first picture of artificial transmutation of elements ever taken. It was made by Patrick Blackett, a student of Lord Rutherford.

> 在某种意义上，可以说物理学家实现了历史上炼金术士们想要转变物质的梦想。

The transformation is typical of many other nuclear transformations studied in experimental physics today. In all transformations of this kind, the incident particle(proton, neutron or alpha particle) penetrates into the nucleus, kicks some other particles out, perhaps remain-ing itself in their place. In all such transformations a new element is formed in the reaction.

Just before the second world war, two German chemists, O. Hahn and F. Strassmann, discovered a different type of nuclear transformation: a heavy nucleus breaks up into two almost equal parts with the liberation of a tremendous amount of energy. In my next slide, you see two uranium fragments flying in opposite directions

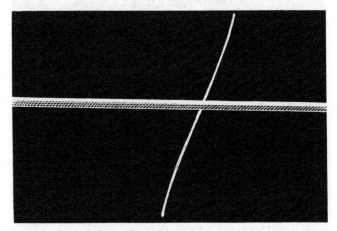

This phenomenon known as nuclear fission

from a thin uranium filament. This phenomenon, known as *nuclear fission*, was noticed first in the case of uranium bombarded by a beam of neutrons. But it was soon found that other elements, also located near the end of the periodic system, possess similar properties. It seems that these heavy nuclei are at the very limit of their stability. The smallest provocation, caused say by a collision with a neutron, is enough to make them break into two—like an oversized, wobbling droplet of water. This instability of heavy nuclei holds the clue as to why there are only 92 elements in nature. Any nucleus heavier than uranium could not exist for any significant period of time without immediately breaking into much smaller fragments—this happening quite spontaneously, without any outside stimulation.

The phenomenon of nuclear fission is interesting from a practical point of view: it can be a source of nuclear power. When they break up, the nuclei emit energy in the form of radiation and fast-moving particles. Among the ejected particles are neutrons. These may go on to cause the fission of neighbouring nuclei. These in turn can lead to yet more neutrons being emitted, yielding yet further fissions—a so-called *chain reaction*. Given enough uranium material—what we call the *critical mass*—the emitted neutrons have such a high probability of hitting other nuclei and causing further fissions, that the process becomes self-perpetuating. Indeed, it can trigger an explosive reaction in which the energy stored inside the nuclei is set free in a fraction of a second. This was the principle used in the first nuclear bombs.

> 新的"炼金术"虽然不靠金子卖钱，却有更大的用途，有时甚至是比金钱更有危害的用途。

The chain reaction does not have to lead to an explos-ion. Under carefully controlled conditions, the process can be contained, leading to a steady, sustained release of energy. This is what happens in nuclear power stations.

The nuclear fission of heavy elements like uranium is not the only way of tapping into the energy of the nucleus. There is a totally different way of doing it. This involves fusing together the lightest elements such as hydrogen to produce heavier ones. This process is known as nuclear fusion. When two light nuclei come into contact they fuse together as do two droplets of water in a saucer. This can happen only at a very high temperature, since the light nuclei approaching each other are kept from coming into contact by the electric repulsion. But when the tempera-ture reaches tens of millions of degrees, electric repulsion is powerless to prevent the contact, and the fusion process starts. The most suitable nuclei for the fusion process are deuterons, i.e. the nuclei of heavy hydrogen

> 极高的温度，是核聚变发生的重要条件。

atoms—deuterium being readily extracted from ocean water.

Now you might be wondering how both fission and fusion can lead to the release of energy. The important point to grasp is that certain combinations of neutrons and protons are more tightly bound than others. Whenever one goes from a more diffuse set-up to one in which the nucleons are more efficiently bound, the excess energy is available to be released. It turns out that large uranium nuclei are rather inefficiently bound, and can be transformed into tighter combinations by splitting up into smaller groupings. At the other end of the periodic table, it is the heavier combinations of nucleons that are the more efficiently bound. A helium nucleus, for instance, consisting of two protons and two neutrons, is exception-ally tightly bound—as we have earlier noted. Thus there is energy available for release if separate nucleons or deuterons can be persuaded to collide and stick together as helium.

And that is where the hydrogen bomb comes in. It is based on the conversion of hydrogen into helium

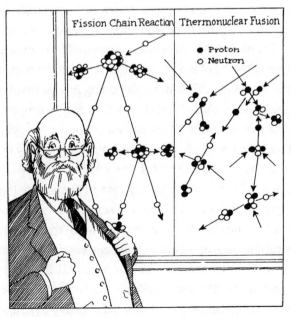

Both fission and fusion can lead to the release of energy

through reactions involving fusion. Much greater outputs of energy are involved here—hence the greater power of hydrogen bombs over the first generation of nuclear weapons based on fission. Unfortunately, it has proved much more difficult to harness the power of the hydrogen bomb for peaceful purposes. Power stations producing energy from nuclear fusion still have a long way to go before achieving commercial viability.

同样不寻常的是，和平利用核聚变能还有很长的路要走，为战争目的而制造氢弹却容易得多。

The Sun, however, has no difficulty in doing this. The continuous conversion of hydrogen into helium is the Sun's main source of energy, and it has succeeded in sustaining this reaction at a steady rate for the past 5000 million years—with another 5000 million years still to go.

In stars more massive than the Sun, higher internal temperatures prevail, and many additional fusion reactions occur. These convert the helium into carbon, the carbon into oxygen, etc. all the way up to iron. Beyond iron no further energy is to be gained from fusion. Instead, as we have noted for the heavier nuclei like uranium, more efficient packing of nucleons, and hence energy release, is to be derived from the opposite process: fission.

13 The Woodcarver
老木雕匠

Arriving home that evening from the lecture, Mr Tompkins found Maud had already retired to bed and was fast asleep. He made himself a drink of hot chocolate and joined her. He sat in bed for a while thinking back over the lecture. He especially recalled the part to do with nuclear bombs. The threat of nuclear annihilation had always disturbed him.

> 这次直接就做梦了。

'This won't do,' he mused. 'I'll be giving myself nightmares if I'm not careful.'

He put down the empty mug, switched off the light, and snuggled up to Maud. Fortunately his dreams were not at all unpleasant... Mr Tompkins found himself in a workshop. At one side was a long wooden bench covered with simple carpenter's tools. On the old-fashioned shelves attached to the wall, he noticed a large number of different wood carvings of strange and unusual shapes. An old, friendly looking man was working at the table. Observing more closely his features, Mr Tompkins was struck by his strong resemblance both to the old man Gepetto in Walt Disney's Pinocchio, and to a portrait of the late Lord Rutherford of Nelson he had seen hanging on the wall of the professor's laboratory.

> 从后文看，梦中的老先生就应是卢瑟福的化身，但肯定不是卢瑟福本人，否则后面他就不该讽刺自己的工作了。

'Excuse me,' ventured Mr Tompkins, 'I couldn't help noticing, but you look a lot like Lord Rutherford—the nu-clear physicist. You aren't by any chance related, are you?'

'Why do you ask?' replied the old man, setting aside the piece of wood he was carving. 'Don't tell me

13 The Woodcarver

you're interested in nuclear physics.'

'Well, as a matter of fact, yes,' said Mr Tompkins, adding diffidently, 'not as an expert, I hasten to say...'

'In that case, you came to just the right place. I make all kinds of nuclei. I'd be glad to show you around my little workshop.'

'You *make* them, did you say?' exclaimed Mr Tompkins.

'Of course. Naturally, it requires some skill—especially in the case of radioactive nuclei. They tend to fall apart before I've had time to paint them.'

涂色，形象的描述。

'*Paint* them?'

'Yes, I use red for the positively charged particles, and cyan (this peacock blue) for the negative ones. Red and cyan are complementary colours; mix them together and they cancel each other out—the mix is colourless.'

'I don't think so,' Mr Tompkins protested mildly. 'Not *colourless*, surely. If I mix red and greeny-blue paint I get...well, a muddy sort of colour.'

The woodcarver smiled. 'Quite right. It's not colour-less if you mix the pigments. But if you look at a mixture of red light and greenyblue light, it produces the sensation of whiteness.'

Mr Tompkins still looked doubtful.

'If you don't believe me,' continued the old man, 'all you have to do is paint one half of a top red and the other half cyan—like this one I have here—and give it a quick spin. See; it looks white—colourless. Anyway, as I was saying, I paint the protons in the atomic nuclei red for their positive charge, and the electrons outside the nucleus cyan for their negative charge. This corresponds to the mutual cancellation of positive and negative electric charges. If the atom is made up of an equal number of positive and negative charges moving rapidly to and fro. it will be electrically neutral and will look white to you. If there are more positive or more negative charges, the whole system will be coloured red

这个梦里对原子核的加工制造实在是太夸张了!

THEY TEND TO FALL APART BEFORE I'VE HAD TIME TO PAINT THEM

or greeny-blue. Simple. Yes?'

Mr Tompkins nodded.

'Now,' continued the woodcarver, showing Mr Tompkins two large wooden boxes standing near the table, 'this is where I keep the materials from which various nuclei can be built. The first box contains *protons*, these red balls here. They are quite stable and keep their colour permanently—unless you scratch it off with a knife, or some-thing. I have much more trouble with the *neutrons* in this other box. They are normally white, or electrically neutral. But they have a strong tendency to turn into red

13 The Woodcarver

protons. As long as the box is closed tight, everything is all right, but as soon as you take one out...Well, see for yourself.'

Opening the box, he took out one of the white bails and placed it on the table. For a while nothing happened. Then, just when Mr Tompkins was about to lose patience, the ball suddenly came alive. Irregular reddish and greenish stripes appeared on its surface, and for a short while the ball looked like one of the coloured glass marbles children like so much. Then the greeny-blue colour became concentrated on one side, and finally separ-ated itself entirely from the ball, forming a brilliant droplet the colour of a peacock, which fell to the floor. At the same instant, a tiny white ball emerged and shot across the room, disappearing into the wall. Meanwhile, the ball itself was now left completely red, no different from any of the red-coloured protons in the first box.

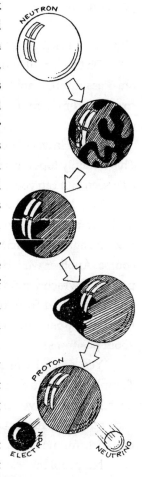

这可真是形象化的描述，可惜在黑白图中看不到色彩。

'Did you see that?' the old man asked, excitedly. 'The white colour of the neutron broke up into red and cyan, and then the whole thing split into three separate particles: this electron,' he said, picking up the ball from the floor. 'You see, it's just an ordinary electron like any other. Then there is the proton on the table (again, a perfectly ordinary proton), and that neutrino.'

'That what?' asked Mr Tompkins looking puzzled. 'Sorry. The last one you mentioned—what was that again?'

> 神奇不可思议的中微子!

> 注意这里用来形容中微子的形容词: slippery.

> 需要能量: 这总是变化的代价。

> 这可是梦中讲梦了。

'A neutrino,' repeated the woodcarver. 'It went over there,' he added, pointing to the distant wall. 'Didn't you notice it?'

'Yes, yes. I saw it,' replied Mr Tompkins hurriedly. 'But where has it gone? I don't see it any more.'

'Oh that's neutrinos for you. Terribly slippery they are. Pass through anything. Closed doors, walls. I could fire one from here right through the Earth and out the other side.'

'Good Heavens!' exclaimed Mr Tompkins. 'How odd. This certainly tops any coloured handkerchief trick I have ever seen. But can you change the colours back again?'

'Yes, I can rub the cyan paint back on to the surface of the red ball (by pushing an electron in) and make it white again, but that would require some energy, of course. Another way to do it would be to scratch the red paint off, which would take some energy too. Then the paint scratched from the surface of the proton would form a red positron—a positive electron. Do you know about positrons?'

'Yes, when I was an electron myself...,' began Mr Tompkins, but checked himself quickly. 'I mean, I have heard that positive and negative electrons annihilate each other whenever they meet,' he said. 'Can you do that trick for me too?'

'No problem,' said the old man. 'But I won't take the trouble to scratch the paint off this proton, as I have a couple of positrons left over from my morning's work.'

Opening one of the drawers, he extracted a tiny bright red ball, and, pressing it firmly between finger and thumb, put it beside the green one on the table. There was a sharp bang, like a fire-cracker exploding, and both balls vanished at once.

'You see?' said the woodcarver, blowing on his slightly burned fingers. 'That's why one cannot use electrons for building nuclei. I tried it once, but gave it up right away. Now I use only protons and neutrons.'

13 The Woodcarver

'But neutrons are unstable too, aren't they?' asked Mr Tompkins, remembering the recent demonstration.

'On their own, yes. But when they're packed tightly in the nucleus, and surrounded by other particles, they become quite stable. Unless,' he added hastily, 'there are too many neutrons in the nucleus—relative to the number of protons. Then they can transform themselves into protons, with the extra paint being emitted from the nucleus in the form of a negative electron. Same way, if there are too many protons, they will change into neutrons, getting rid of their unwanted red paint in the form of a positive electron. Such adjustments we call beta transformations. 'Beta' is the old name given to the electrons emitted from such radioactive decays.'

'Do you use any glue, in making the nuclei?' asked Mr Tompkins with interest.

> 胶水！又是形象的说法。

'Don't need any,' answered the old man. 'These particles, you see, stick to each other by themselves as soon as you bring them into contact. You can try it yourself if you want to.' Following this advice, Mr Tompkins took one proton and one neutron in each hand, and brought them together carefully. At once he felt a strong pull, and looking at the particles he noticed an extremely strange phenomenon. The particles were exchanging colour, becoming alternately red and white. It seemed as if the red paint were 'jumping' from the ball in his right hand to the one in his left hand, and back again. This twinkling of colour was so fast that the two balls seemed to be connected by a pinkish band along which the colouring was oscillating to and fro.

'This is what my friends the theoretical physicists call the "exchange phenomenon", said the old master, chuckling at Mr Tompkins's surprise. 'Both balls want to be red, or to have the electric charge, if you want to put it that way, and as they cannot have it simultaneously, they pull it to and fro alternately. Neither wants to give up, and so they stick together until you separate them by force. Now I can show you how simple it is to make any nucleus you want. What shall it be?'

'Gold,' said Mr Tompkins, remembering the ambition of the medieval alchemists.

'Gold? Let us see,' murmured the woodcarver, turning to a large chart hanging on the wall. 'The nucleus of gold weighs one hundred and ninety-seven units, and carries seventy-nine positive electric charges. That means I have to take seventy-nine protons and add one hundred and eighteen neutrons to get the mass correct.'

Counting out the proper number of particles, he put them into a tall cylindrical vessel and covered it all with a heavy wooden piston. Then, with all his strength, he pushed the piston down.

'I have to do this,' he explained to Mr Tompkins, 'because of the strong electric repulsion between the positively charged protons. Once this repulsion is overcome by the pressure of the piston, the protons and the neutrons will stick together because of their mutual exchange forces, and will form the desired nucleus.'

Pressing the piston in as far as it would go, he took it out again and quickly turned the cylindrical vessel upside down. A glittering pinkish ball rolled out on the table. Watching it closely, Mr Tompkins noticed that the pinkish colour was due to an interplay of red and white flashes among the rapidly moving particles.

'How beautiful!' he exclaimed. 'so this is an atom of gold!'

'Not an atom yet, only the atomic nucleus,' the old man corrected him. 'To complete the atom we have to add the proper number of electrons to neutralize the positive charge of the nucleus, and make the customary electronic shell around it. But that's easy, and the nucleus itself will catch its electrons as soon as there are some around.'

'It's funny my father-in-law never mentioned that one could make gold so simply,' mused Mr Tompkins.

'Oh your father-in-law, and those other nuclear physicists!' murmured the old man with a twinkle in his eye. 'Yes, of course, they can transform one element into another to a limited extent. But the quantities they get are so small they can hardly be seen. Let me show you how they do it.'

With that, he took up a proton, and threw it with considerable force against the gold nucleus lying on the table. Nearing the outside of the nucleus, the proton slowed down a little, hesitated a moment and then pl-unged inside it. Having swallowed the proton, the nucleus shivered for a short time as though in a high fever and then a small part of it broke off with a crack.

'You see,' he said, picking up the fragment. 'This is what they call an alpha particle. If you inspect it closely you'll notice that it consists of two protons and two neutrons. Such particles are usually ejected from the heavy nuclei of the so-called radioactive elements. But one can also kick them out of ordinary stable nuclei if one hits them hard enough. Note that the larger fragment left on the table is no longer a gold nucleus; it gained one positive charge, but lost two when the alpha was emitted, meaning it has lost one positive charge overall. It's now a nucleus of platinum, the preceding element in the periodic table. In some cases, however, the proton which enters the nucleus will not cause it to split in two parts, and as a result you will get the nucleus that follows gold in the table, i.e. the nucleus of mercury. Combining these and similar processes one can actually transform any given element into any other.'

'So why don't physicists turn lots of the common elements, such as lead, into the more valuable ones—gold, say?' asked Mr Tompkins.

'Because firing projectiles at nuclei is not very effective. First of all they cannot aim their projectiles the way I can in my piston tube. Only one in several thousand shots actually hits the nucleus. Second, even in the case of a direct hit, the projectile is very likely to bounce off the nucleus instead of penetrating into the interior. Did you

> 虽然物理学家有了"点石成金"的本事，可要靠此发财还是不太容易。自然规律决定了这一点。

notice how when I threw the proton into the gold nucleus, it hesitated a bit before going in. For a moment, I thought it was going to be thrown back—as so often happens.'

'What's there to prevent the projectiles from going in?' asked Mr Tompkins.

'You don't need me to tell you *that*, surely,' replied the old man somewhat reproachfully. 'Think! Both the nucleus and the bombarding proton carry positive charges. The repulsive force between these charges forms a kind of barrier. It's not easy for the proton to get across this barrier. If the bombarding proton manages to penetrate the nuclear fortress, it is only because it uses something like the Trojan horse technique; they go *through* the nuclear walls, not over them. It's something they can do as waves, rather than particles.'

Mr Tompkins was about to confess that he did not understand what on Earth the old man was on about, when it gradually dawned on him that he probably *did* understand!

'I once saw a funny snooker game,' he said. 'There was this ball. It started off inside the wooden triangle. Then all of a sudden it was outside. It sort of 'leaked' through the wooden barrier. It got me worrying about the possibility of tigers leaking out of their cages. Do you think that's the same kind of thing we've got here—only now instead of a snooker ball, or tiger, leaking out, we have a proton leaking in?'

'Sounds like it to me. But to tell the truth, theory was never my strong point. I'm a practical man myself. Still, it does seem pretty obvious that these nuclear particles—given that they are made out of quantum material—are going to leak through obstacles ordinarily considered impenetrable.'

He looked hard at Mr Tompkins. 'Those snooker balls,' he added. 'They weren't by any chance real quantum ivory snooker balls?'

'Yes' replied Mr Tompkins. 'I understand they were

13   The Woodcarver

made from the tusks of quantum elephants.'

'Well, such is life,' said the old man sadly. 'They use such rare materials just for games, and I have to carve protons and neutrons, the basic particles of the entire universe, out of plain quantum oak! But,' he continued, trying to hide his disappointment, 'my poor wooden toys are probably just as good as all those expensive ivory creations. Here, let me show you how neatly they can pass through any kind of barrier.'

> 这就是生活！这就是人生！这就是……

Climbing onto the bench, he took from the top shelf a carved object, which Mr Tompkins at first took to be the model of a volcano.

'What you see here,' the woodcarver continued, gently blowing off the dust, 'is a model of the typical barrier of repulsive forces surrounding an atomic nucleus. The outer slopes correspond to the electric repulsion between the charges, and the crater to the cohesive forces which make the nuclear particles stick together. If I now flip a ball up the slope, but not hard enough to bring it over the crest, you would naturally suppose that it would roll back again. But see what actually happens...'

He gave the ball a slight flip. After rising about half way up the slope, it rolled back down again to the table surface.

> 再次演示"崂山道士"的把戏。

'So?' remarked Mr Tompkins, unimpressed.

'Wait,' said the woodcarver quietly. 'You can't expect success at the first attempt.'

Again, he sent the ball up the slope; but again it failed. It was not until the third time he struck it lucky: the ball suddenly disappeared just when it was about half-way up the slope.

'Hah!' the woodcarver declared triumphantly, with the air of a magician. 'Abracadabra! The famous vanishing trick. What do you make of that? Where's it gone?'

'Into the crater?' enquired Mr Tompkins uncertainly.

'That's what I think,' agreed the old man. 'Let's see...' He peered into the top of the crater. 'Yes, that's

Again he sent the ball up the slope

exactly where it is,' he added, picking out the ball with his fingers.

'Now, let's see the reverse,' he suggested. 'Let's see if the ball can get out of the crater without rolling over the top.'

Having carefully replaced the ball in the hole, they waited. For a while nothing happened. Mr Tompkins could hear the rumbling of the ball rolling to and fro in the crater. Then, as if by a miracle, the ball suddenly appeared on the outside, about halfway up the slope, and quietly rolled down to the table.

'What you see this time is a pretty good representation of what happens in radioactive alpha decay,' said the woodcarver. Putting the model back on its shelf, he added. 'Sometimes these electric barriers are so "transparent" the particle escapes in a small fraction of a second; sometimes they are so "opaque" it takes many billion years—as for example, in the case of the uranium nucleus.'

'But why aren't *all* nuclei radioactive?' asked Mr Tompkins.

'Because in most nuclei the floor of the crater is below the outer level, and only in the heaviest known nuclei is the floor sufficiently elevated to make such an escape

可是,《聊斋》中的"崂山道士"带有放射性吗?

13 The Woodcarver

possible.'

The woodcarver looked up at the clock on the wall. 'Goodness, is that the time. I should be shutting up shop. Do you mind...'

'Oh I'm sorry. I didn't mean to take up your time like this,' apologised Mr Tompkins. 'But it has been most interesting. Just one more thing though—if I may?'

'Yes?'

'You said that firing projectiles at nuclei was a very inefficient way of transforming base elements into more valuable ones...'

The woodcarver smiled, 'Still hoping to make your fortune through nuclear physics?'

Mr Tompkins shifted uncomfortably, but continued. 'And yet you seem to have no difficulty with your clever device over there.' He pointed in the direction of the tube and piston contraption. 'So, I was wondering...'

The woodcarver smiled. 'Clever it is; real it isn't. That's the problem. No, you'll just have to accept that the conversion of base, metals into gold—commercially speaking—is a pipe dream. It's time for you to wake up, I'm afraid.'

'What a shame,' thought Mr Tompkins, disconsolately.

'I said, it's time to wake up!'

This time it wasn't the woodcarver speaking. It was Maud.

> 想发财的梦总是容易醒。

195

14 Holes in Nothing
虚空中的空穴

> 又是开门见山。

> 说起反物质，不妨参读一下那位因《达·芬奇密码》而走红的美国作家丹·布朗的另一本小说：《天使与魔鬼》，也很好读，也涉及反物质！

Ladies and gentlemen:

Tonight we tackle a particularly fascinating topic: antimatter.

The first example of an antiparticle was the positron— the positive electron of which I have spoken in past lectures. It is noteworthy that the existence of this new kind of particle was predicted on the basis of purely theoretical considerations. This was several years before it was actually detected. In fact, the experimental discovery was much helped by knowing in advance what its main properties were expected to be.

The honour of having made this prediction belongs to the British physicist, Paul Dirac. Using Einstein's theory of relativity, and incorporating the requirements of quantum theory, he deduced a formula for the energy, E, of an electron. Towards the end of the calculation he arrived at an expression for E^2. So, the last step consisted in taking the square root of this expression to find the formula corresponding to E itself. As is usual, when you take a square root, there are two possible answers: one positive, the other negative. (For example, the square root of 4 can be +2 or — 2.) In solving physical problems, it is customary to ignore the negative solution as being "unphysical"; in other words, it is taken to be a mere quirk of the mathematics of no significance. In this particular case, the negative solution would have corresponded to an electron having negative energy. Bearing in mind that, according to relativity theory, matter is a form of energy in itself, a negative energy electron would imply that it had *negative mass*. This would be really weird! Exert a pull on such a particle, and it would move away from you; try to push it away, and it would come towards you—the very opposite of what is expected of 'sensible' positive-massl particles. Surely, it might be thought, this is reason enough for regarding the negative solution to the equation as 'unphysical'. Ignore it !

The genius of Dirac lay in the way he did *not* take that line. In addition to the infinite number of different positive energy quantum states open to an electron, he took the negative solution to imply that there was also an infinite number of negative-energy states open to it. The problem was that as soon as an electron found itself in one of the latter, it would have to exhibit the behaviour characteristic of negative mass—and, of course, no such thing has been observed. So, where *are* these hypothetical, bizarre negative-mass electrons?

One might at first try to wriggle out of this by saying that it just so happens electrons avoid those particular states; for some reason they are left permanently empty. But this will not do. We already know that with the quantum energy states available to an electron in an atom, the electrons have a natural tendency to radiate away their energy and drop down to the lowest available energy state (consistent with that state not already being occupied by another electron—in accordance with the Pauli exclusion principle). That being so, we would expect *all* electrons in time to drop down from the higher positive energy states to the lower negative energy states. They should *all* behave badly!

The solution Dirac proposed was the strangest possible. He claimed that the reason why the electrons we know of do not drop down into the negative energy states, is that all those states are already full; the infinite number of negative energy states are filled by an infinite number of negative-mass electrons! If that is really so, why do we not see them? Precisely because there are so many of them—they form a perfect continuum. The electrons are

科学经常是在这种许多人认为天经地义地没意义的地方取得重大发现。

多么大胆的想法！

there in the 'vacuum' in a completely regular and uniform distribution.

A perfect continuum is undetectable. You can't point to it and say "There it is." It is everywhere. There is not 'more' of it in one region than another. As you move through it there is no build up of its density in front of you, leaving a 'gap' behind you—as is the case when a car moves through air, or a fish through water. Thus there is no resistance to motion...'

> 听这么专业的讲座,能不困吗?

By this point in the lecture, Mr Tompkins felt his brain reeling. A vacuum—total emptiness—being completely full of something! It's all around you and within you, but you don't notice it!

He got to day-dreaming about what it would be like to be a fish, spending all one's life in water. He felt a gentle warm breeze off the sea—an ideal day to take a dip in the gently rolling blue waves. So it was he joined the fish in their watery world. Though he was a good swimmer, on this occasion he found himself sinking deeper and deeper beneath the surface. But strangely enough, he did not notice the lack of air, and felt quite comfortable. Maybe, he thought, this is the effect of a special recessive mutation. He recalled how, according to palaeontologists, life originated in the ocean, with the first fish pioneer to get out on dry land being similar to the so-called *lungfish*. It crawled out onto a beach, walking on its fins. According to biologists, these first lungfish gradually evolved into land-dwelling animals, like mice, cats and humans. But some of them, like whales and dolphins, after learning of all the troubles of life on dry land, returned to the ocean. Getting back to the water, they retained the qualities acquired during their struggle on the land, and remained mammals, the females bearing their progeny inside their bodies instead of just dropping eggs and having them fertilised later by the males.

It was while he was lazily swimming about and

14 Holes in Nothing

having such thoughts that he came upon a strange couple; one was a man bearing an uncanny resemblance to Paul Dirac (Mr Tompkins recognised him from a slide photograph the professor had briefly flashed up in his lecture), the other was a dolphin. They were deep in conversation. This did not strike Mr Tompkins as being in any way odd; dolphins he remembered were very intelligent.

My 'water' is frictionless and uniform everywhere

'Look here, Paul,' the dolphin was saying, 'you contend that we are not in a vacuum but in a material medium formed by particles with negative mass. As far as I am concerned, water is no different from empty space; it is completely uniform and I can move freely through it in all directions. In our dolphin community, a legend has been handed down from our pre-pre-pre-pre-predecessors that dry land is quite different. There are mountains and canyons which one cannot cross without effort. But here in water we can move in any direction we choose.'

'You're right, my friend,' answered Dirac. 'Water exerts friction on the surface of your body; this helps you get a 'grip' on the water. You are able to build up pressure differentials in the water by the way you move your fins and tail. This helps you to swim—to move about. But

海豚据说很聪明，但却难以超越来自直接经验的感受。

if water had no friction, and if there were no pressure gradients because the water was perfectly uniform everywhere, then you would be as helpless as an astronaut who had run out of rocket fuel.

'My "water", which is formed by electrons with negative masses, is quite different. It is completely frictionless and uniform everywhere, and is therefore unobservable. Another difference is that not a single electron can be added to it. This is because of the Pauli exclusion principle (which you recall forbids more than two electrons with opposite spins from occupying the same quantum state). In my "water" all the possible quantum levels are already occupied. Any extra electron would have to remain *above* the ocean's surface. This in turn means it has the normal positive mass and behaves like a normal electron.'

'But,' persisted the dolphin, 'if your ocean is unobservable because of its continuity and absence of friction, what is the sense of talking about it?'

'Well,' replied Dirac, 'assume that some external force lifted one of the electrons with negative mass from the depth of the ocean to above its surface. In this case the number of observable electrons will increase by one. But not only that, the empty hole in the ocean from which the electron was removed will now be observable.'

'It would act like one of the bubbles down here,' suggested the dolphin. 'Like that one over there,' he continued, pointing to a bubble lazily emerging from the depths and making its way to the surface.

'Exactly.' agreed Dirac. 'In my world we would not only see the electron that had been knocked up into the positive-energy state, we would also see the hole left behind in the vacuum. The hole is the *absence* of whatever was there before. Thus for instance, the original electron had a negative electric charge; the *absence* of that negative charge from a uniform distribution would be perceived as the presence of an equal amount of positive charge. The absence of its negative mass would moreover be

> 海豚还是坚持经验观察的意义。

perceived as a positive mass—the same size mass as the original electron, only positive. In other words the hole behaves as a perfectly normal, sensible particle. It behaves like an electron, except it carries positive charge rather than the usual negative charge. It's what we call the positron. So that's what we have to look out for: *pair production*—the simultaneous appearance of an electron and a positron at the same point in space.'

'A very clever theorys' remarked the dolphin. 'But is it true... ? '

聪明的海豚虽然还是怀疑，却也不得不感叹理论家的智慧。

'Next slide.' The familiar commanding voice of the professor broke into Mr Tompkin's reverie:

As I was saying, the only way to detect the continuum would be if you could somehow *disturb* it. That way it would no longer be a perfect continuum. If you knocked a hole in it, you would then be able to say 'The continuum is everywhere—except *there*.' That, ladies and gentlemen, is precisely what Dirac suggested: you knock a hole in empty space. And this picture shows it being done!

梦被吵醒，文字变难懂了。

It is a bubble chamber photograph. I should perhaps explain that a bubble chamber is a particle detector somewhat like the Wilson cloud chamber—but 'turned inside out'. It was invented by the American physicist Donald Glaser, securing for him the Nobel Prize in 1960. According to his story, he was once sitting in a bar, gloomily watching bubbles rising in the beer bottle which stood in front of him. He suddenly thought, if Wilson could study liquid droplets in a gas, why couldn't he do better by studying gas bubbles in a liquid? Instead of expanding a gas to create a supersaturated vapour that tries to condense, why not release the pressure on a liquid so that it becomes superheated and tries to boil? And that is what the bubble chamber does: it marks out the trails of charged sub-atomic particles with trails of bubbles.

This particular slide shows the production of two

electron-positron pairs. A charged particle enters at the bottom of the picture. It undergoes an interaction at the point where you see the kink. From this interaction there emerges not only the charged particle leaving the track veering off to the right, but also a neutral particle which promptly changes into two high energy gamma rays. You cannot see either this second particle or the gamma rays it produces because they are electrically neutral and so leave no trail of bubbles. But then each of the gammas gives rise to an electron-positron pair—the V-shaped confi-guration of tracks at the top of the picture. Note how both the "V's" are pointing back to the vicinity of the original interaction.

> 注意，对这里就照片上看不见部分的说明，是基于理论解释的，而且是基于以前更常识性的理论。

Note also that all the tracks are systematically curved to one side or the other. This is because there is a powerful magnetic field operating over the whole area of the chamber, directed along our line of sight. This causes negatively charged moving particles to curve clockwise in the photograph, and positively charged particles to curve in the opposite sense. Having told you that, you should now be able to distinguish the positron from the electron in each pair. Incidentally, the reason why some tracks are more curved than others is that the amount of bending depends on the particle's momentum; the smaller the momentum, the greater the curvature. As you will begin to appreciate, a bubble chamber picture is full of clues as to what is going on there!

Pointing back to the vicinity of the original interaction

Now you have seen how to knock holes in a vacuum,

14 Holes in Nothing

you will doubtless be wondering what happens to them after that...

In point of fact, that was not what Mr Tompkins was wondering about. His thoughts had already gone back to the time when he was himself an electron. He remem-bered with a shiver having to dodge the predatory positron. But the professor continued:

如果失去联想，汤普金斯先生将会怎样？

...The positron continues behaving like a normal particle—until that is it meets up with an ordinary negatively charged electron. The electron promptly falls into the hole, filling it up. The continuum is restored, and both the electron and the hole disappear; we call this the *mutual annihilation* of a positive and a negative electron. The energy set free in the fall is emitted in the form of photons.

One general point I ought to make is that I have been referring to negative electrons as the overflow of Dirac's ocean and to positrons as the holes in it. One can, however, reverse the point of view and consider ordinary electrons as the holes, giving to positrons the role of thrown-out particles. Both pictures are absolutely equivalent from the physical as well as the mathematical point of view.

Next, electrons are not unique in having an *antiparticle*—as we call the positron. Corresponding to the proton there is an *antiproton*. As you would expect, it has exactly the same mass as the proton, but the opposite electric charge; in other words, antiprotons are negatively charged. The antiproton can be regarded as a hole in another type of continuum—one consisting this time of an infinite number of negative-mass protons. Indeed, all particles have antiparticles. The vacuum really does contain a very great deal!

One question that might have occurred to you is why the world we know has such a preponderance of matter as distinct from antimatter. This extremely

这里的问题已很有哲学味了。

interesting question is a very hard one to answer. In fact, since atoms built by positive electrons surrounding negative nuclei would have exactly the same optical properties as ordinary atoms, there is no way to decide from any spectroscopic observation whether distant stars are made of our type of matter or its opposite. For all we know, it is quite possible that the material forming, let us say, the Great Andromeda Nebula is of this topsy-turvy type. The only way to prove it would be to get hold of a piece of that material and see whether or not it is annihilated by contact with terrestrial materials. (There would, of course, be a terrible explosion!)

In point of fact we do not have to embark on such a dangerous mission. It is quite common to observe galaxies colliding with each other. If one were made of matter and the other of antimatter, the amount of energy released as the electrons of one galaxy annihilated the positrons of the other would be spectacular indeed. Observations reveal nothing to suggest that this is happening. Thus it seems fairly safe to assume that almost all the matter of the Universe is of one type only. It is not the case that half the galaxies are matter and the other half antimatter.

Recently there have been suggestions that there might have been equal numbers of the two types of matter at the very beginning of the Universe. But then, in the course of the development of the Big Bang, interactions tended to favour one rather than the other. It was this subsequent behaviour that led to the present imbalance. This, however, is but a tentative suggestion at the moment.

> 这也是在给读丹·布朗的小说做准备知识的科普。

> 可是，在科学发展的绝大多数阶段，都充满了 tentative suggestion。

15 Visiting the 'Atom Smasher'

参观"原子粉碎机"

Mr Tompkins could hardly contain his excitement. The professor had arranged for a party of his students to visit one of the world's foremost high-energy physics laboratories. They were about to see an atom smasher!

In the weeks preceding, they had each been issued with a brochure. Mr Tompkins had dutifully read it from cover to cover. Not that it seemed to make much sense. His mind was a complete blur: jumbled-up ideas about quarks, gluons, strangeness, energy changing into matter, and grand unifying theories that explained everythingthough not to him.

On arrival at the Visitors' Centre, they were ushered into a waiting room. They did not have long to wait before their guide came bustling in. A bright-eyed, earnest-looking young woman in her mid-twenties, she welcomed them and introduced herself as Dr Hanson, a member of one of the research teams.

'Before we go over to the accelerator, I would just like to say a few words about what we do here.'

A man tentatively raised a hand.

'Yes?' Dr Hanson asked. 'You have a question?'

'You said "accelerator". What about the atom smasher? Aren't we going to see that as well?'

The guide gave a slight grimace. 'That's what I was talking about. The machine—the accelerator—it's what

> 这一章,及随后的两章,都是在伽莫夫的原作出版了几十年后,由另一位"合作者"(之所以打引号,是因为在他写作时伽莫夫早已去世了)新写的。他虽然努力模仿伽莫夫的风格,但读者还是可以体会出其间的差异。

> 这倒确实像一次游览,而且导游是不可缺少的。

A member of one of the research teams

newspapers call an "atom smasher". But that's not what we call it. It's misleading. After all, if you simply want to smash an atom, you knock some of its electrons off. Easy. Even smashing up the atom's nucleus is relatively easy—at least, compared with what we do here. So we call it a "particle accelerator".

'Any further questions? Please feel free...' She looked around the audience. There being no response, she continued.

'Right, then. The overall aim is try to understand the tiniest bits of matter and what holds them together. As you doubtless know, matter is made of molecules, molecules of atoms, and an atom is composed of a nucleus and electrons. Electrons are thought to be elementary—in other words they are not made up from even more fundamental constituents. But this is not the case for the nucleus; the nucleus is made of protons and neutrons. I take it all this is familiar?'

The audience nodded.

'So, it's pretty obvious what the next question will be...'

'What are protons and neutrons made of ?' a lady suggested.

'Exactly. And how do you suggest we find out?'

'Smash them up?' she ventured.

> 这些知识对汤普金斯先生已是小菜一碟了。

15 Visiting the 'Atom Smasher'

'Yes indeed. That seems the right approach. We find out about the structure of molecules, then atoms, and then nuclei by hurling projectiles at them and breaking them apart. So that's what we start out trying to do; we accelerate particles—either protons or electrons—to high energy and make them collide with protons. That way we hope to split up the proton into its constituent parts.

'And what happens?' she continued. 'Does the pro-ton break up? No. Regardless of how energetic the projectile, the proton *never* splits up. Instead, something else happens—something quite remarkable: The collision leads to the creation of new particles—particles that weren't there to begin with.

'For example, collide two protons and you might end up with two protons plus an additional particle, a so-called pion or π particle. It has a mass 273.3 times the mass of an electron, or 273.3 m_e. We write it like this...'

Dr Hanson moved to a flip chart and wrote:

$$p+p \rightarrow p+p+\pi$$

An elderly man immediately raised his arm.

'But surely that's not allowed,' he declared with a frown. 'It's a long time since I did physics at school, but one thing I do remember: Matter can neither be created nor destroyed.'

'I'm afraid I have to tell you that the one thing you learned at school is wrong!' Dr Hanson announced, causing a ripple of laughter.

现在教中学生学物理时已经不会再这样讲了。

'Well, not *entirely* wrong, I suppose,' she added hastily. 'We cannot create matter from *nothing*. That still applies. No, we create matter from energy. It's a possibility allowed by Einstein's famous equation,

$$E=mc^2$$

I take it you've come across this before?'

The students glanced around at each other uncertainly.

这个著名的公式是狭义相对论的推论。

'I'm sure it's something we've all *heard* of,' Mr Tompkins volunteered. 'But I'm not sure we've covered it in our lectures yet.'

'Well, it's a consequence of Einstein's relativity theory,' Dr Hanson explained, 'According to Ein-stein, it is impossible to accelerate a particle faster than the speed of light. One way of understanding this is to think of the mass going up. As the particle goes faster, its mass increases, making further acceleration more difficult.'

'We've covered *that*,' said Mr Tompkins hopefully.

'Oh excellent,' she replied. 'Well in that case, all you have to recognise is that the accelerating particle is getting not only more massive, but also more energetic. The equation $E=mc^2$ means, in effect, that energy, E, has a mass, m, associated with it. (c is the speed of light and is included in order to allow us to write mass in the same units as energy.) So, as the particle accelerates and takes up more energy, it must also take on board whatever mass goes with that energy. And that's why the particle seems to get heavier. The extra mass is due to the extra energy it now has.'

'But I don't understand,' the elderly man persisted. 'You say the extra mass comes with the extra energy. But the particle already had mass when it was stationary-when it had *no* energy.'

'Good point. What we have to remember is that en-ergy comes in different forms: heat energy, kinetic energy of motion, electromagnetic energy, gravitational potential energy,etc. The fact that a stationary particle has mass shows that *matter itself* is a form of energy: "locked-up" or "congealed" energy. The mass of a stationary particle is the mass of its locked-up energy.

'Now. what happened in this collision is that some of the initial kinetic energy of the projectile got transformed into locked-up energy—the locked-up en-ergy of the new pion. We still have exactly the same amo-unt of energy—and of mass—after the collision as befo-

> 注意，这很像一个警句："物质本身就是能量的一种形式！"

15 Visiting the 'Atom Smasher'

re, but now some of the energy is in a different form. OK?'

Everyone nodded.

'Right. So we've created a pion. Now we repeat the experiment. We look at lots and lots of collisions. What we find is that we cannot create new particles of *any* mass: 273.3 m_e yes, but never 274 m_e or 275m_e, say. There are heavier particles—but they only occur at certain allowed masses. For example, there is a K particle with a mass of 966 m_e, in other words, about half the mass of a proton. And there are particles heavier than the proton, such as the Λ (lambda) at 2, 183 m_e. In fact, there are now over 200 known particles, together with their antiparticles. We expect the number to be unlimited. What we can make depends on how much energy is available in the collision. The more energy, the heavier the particle we can produce.

'OK. Having created these new particles, we take a look at them ; we examine their properties. This is not to say we've lost interest in our first question: What is a proton made of? Certainly not. But it turns out that the key to understanding the structure of the proton lies in the study of these new particles—not in the attempt to break the proton down into its constituent parts. The point is that all these new particles are close cousins to the proton. You know how you can sometimes learn about a person by studying their family background. The same applies here; we can learn about the structure of our familiar proton, and neutron, by taking a look at their relatives.

'And what do we find? Well, as you would expect, the new particles are characterised by the normal properties: mass, momentum, energy, spin angular momentum, and electric charge. But in addition to these, they also have *new* properties—properties the proton and neutron do not have. Properties with names like "strangeness" and "charm". Incidentally, don't be fooled by the whimsical nature of these names; each property has a

> 物理与人理的相通之处!

strict scientific definition.'

Someone in the audience raised a hand. 'What do you mean: "a new property"? What kind of property are we talking about? How do you recognise it?'

'Good question,' mused Dr Hanson. She paused for a moment.

'Yes, let me try and put it this way. I'll start with a familiar property. Take a look at the following reaction producing an uncharged pion, or pi zero particle:

$$p^+ + p^+ \rightarrow p^+ + p^+ + \pi^0 \qquad (21)$$

Here the superscript refers to the electric charge carried by the particle. Normally we don't bother to write a+above the p because everyone knows a proton has one unit of positive charge. But for reasons that will become clear later, I wish to spell it out. Here are two more reactions, one producing a pi minus, the other a pi zero:

$$p^+ + n^0 \rightarrow p^+ + p^+ + \pi^- \qquad (ii)$$
$$\pi^- + p^+ \rightarrow n^0 + \pi^0 \qquad (iii)$$

where the symbol n^0 refers to a neutron. All three of these reactions happen. The following does *not* happen:

$$p^+ + p^+ \not\rightarrow p^+ + p^+ + \pi^- \qquad (iv)$$

Now, why do you think that's the case? Why do the first three happen, but the fourth one never does?'

> 这里的算术并不复杂。

'Has it anything to do with the electric charges being wrong?' one of the the younger students asked. 'With the fourth reaction you have two positive charges on the left, and two positives and a negative on the right. They don't balance.'

> 守恒，是物理学中重要的基本概念。

'Exactly. Electric charge is a property of matter, and it has to be conserved. The net charge before the reaction must equal the net charge afterwards—and with the fourth, it doesn't. OK, that's pretty straightforward. But now take a look at this reaction. It involves two of the new particles, the lambda zero and the kay plus:

$$\pi^+ + n^0 \rightarrow \Lambda^0 + K^+ \qquad (v)$$

15 Visiting the 'Atom Smasher'

It is a reaction that is observed to happen. Contrast that with this next one which *never* happens:

$$\pi^+ + n^0 \nrightarrow \Lambda^0 + K^+ + n^0 \qquad \text{(vi)}$$

If you want to produce *that* combination of final particles, you must begin differently:

$$p^+ + n^0 \rightarrow \Lambda^0 + K^+ + n^0 \qquad \text{(vii)}$$

But if you start off with that initial combination, you now find the following won't happen:

$$p^+ + n^0 \nrightarrow \Lambda^0 + K^+ \qquad \text{(viii)}$$

And that's despite the fact that energy-wise it ought to be easier to produce ($\Lambda^0 + K^+$) than ($\Lambda^0 + K^+ + n^0$). So, the question is: What stops reactions (vi) and (viii) from hap-pening?'

She scanned the students' faces. 'Is it anything to do with electric charge conservation this time?'

They shook their heads.

'No. It can't be that,' she said. 'The electric charge balances. So. Any ideas?'

They all looked blank.

'OK. It's at this point we introduce the idea of there being a new property. We call it *baryon number*. The name comes from the Greek word meaning "heavy". We denote it by the letter B. We assign the following values to the particles:

n^0, p^+, Λ^0 all have $B = +1$ unit

π^0, π^+, π^-, and K^+ have $B = 0$

The first group of particles we call "baryons", and the second "mesons"—from the Greek meaning 'middle'. (I should perhaps mention that there are yet other particles, such as electrons, which are light—the so-called "lep-tons".)

'Right, now. Having assigned the B values, we propose that B is conserved: The total amount of baryon number before and after the collision has to be the same. So, with that in mind, I'd like you to take a look at those reactions again. Check that the ones that happen are those that conserve B, while the ones that do not, fail to

重子数，粒子的新性质，幸好这样的东西我们在宏观世界中不用去管它。

引入重子数，就又增加了一种守恒量。

conserve B.'

After a minute or two of concentrated adding and subtracting, the students began to nod, murmuring their agreement.

'Good. It's the failure to conserve B that is *responsible* for those reactions being disallowed. The non-occurrence of those reactions tells us there is a new property, B. What's more, we have learned something about that property: it has to be conserved in collisions—just like electric charge, or energy, or momentum, etc.'

The students were obviously happy with this explanation. Not so Mr Tompkins. He sat there, arms folded, a sceptical look on his face. Dr Hanson noticed.

'Something wrong?' she enquired. 'You have a question?'

'Not exactly a question,' he replied. 'More a comment. Frankly, I'm not convinced. In fact—if you don't mind me s aying so—I think it's all a bit of a fiddle!'

'A fiddle?' she asked in some confusion. 'I don't...Sorry. What are you saying...?'

> 这次汤普金斯真的提了个聪明的好问题，对一个外行来说，实在是不容易。

'The values of the baryon numbers of those particles. Where did you get them from? I reckon you chose them precisely in order to get the results you wanted. You arranged for them to have those values so that the right reactions went, and the others didn't.'

Mr Tompkins' fellow students stared at him in surprise. How dare he? But the tension was quickly resolved; Dr Hanson broke into laughter.

'Very good,' she said. 'Absolutely right. That is how we find out what baryon numbers to assign. We look at reactions that happen, and those that do not, and we make the assignments to fit.'

'But there is more to it than that. If there weren't, it would be a waste of time. The point is this: Having used up a handful of reactions to find out what the particle assignments should be, we can then go on to make predictions about what *other* reactions can and cannot

15 Visiting the 'Atom Smasher'

happen—hundreds and hundreds of predictions.'

Mr Tompkins still looked unconvinced.

'Let me put it like this,' she added. 'One day, a research team announces a big discovery. They have found a new negatively charged particle. They call it the X^-. It was found in the reaction

$$p^+ + n^0 \rightarrow p^+ + p^+ + n^0 + X^- \qquad \text{(ix)}$$

What is its B?'

After some hurried arithmetic, the students started murmuring 'Minus 1?'

'That's right. The total B on the left is +2, whereas on the right we have two protons and a neutron, giving $B=+3$. So, to balance up the two sides, X^- must have $B=-1$, OK, we've "used up" that reaction in order to find out what the value of B is. That's "the fiddle" part,' she said, looking meaningfully in the direction of Mr Tompkins.' The researchers now go on to claim that the X^- particle, directly after it was produced, went on to give the following reaction:

$$X^- + p^+ \rightarrow p^+ + p^+ + \pi^- + \pi^- \qquad \text{(x)}$$

Are you happy with that?'

The students nodded automatically. But then, following a whispered conversation, a few began tentatively shaking their heads.

'What's the matter?' Dr Hanson asked of them. 'You don't believe they've got it right?'

Further discussion. Then one of them explained that, if the B of the X^- really was -1, as they had earlier concluded, then the total B before and after this new reaction did not balance. That meant the reaction could *not* have happened.

这些学生们也挺聪明的。

'Well done! Quite right. They were just kidding! What the X^- really did was this:

$$X^- + p^+ \rightarrow \pi^- + \pi^- + \pi^+ + \pi^+ + \pi^0 \qquad \text{(xi)}$$

This, you can check out, *does* balance. So, what this means is that you have used the baryon number idea to make a *prediction*—the prediction that reaction (x) cannot happen. That's the power of the baryon number idea.' Turning to Mr Tompkins, she asked, 'Satisfied, now?'

He grinned and nodded his assent.

'In fact,' she continued, 'the X^- is an antiproton—us-ually represented by $\bar{p}$. The antiproton has the same mass as the proton but opposite electric charge and B. Reaction (xi) is a typical way in which a proton and an antiproton annihilate each other.

'OK. Now we're getting the idea, let us try the follow-ing reaction. It *never* happens:

$$K^+ + n^0 \nrightarrow \pi^+ + \Lambda^0 \qquad (xii)$$

If you check out the electric charge and B number totals on both sides, they both tally all right. But, as I say, this reaction never happens. Why do you think that might be?'

'There's another property around?' suggested Maud.

又一种新的性质——奇异数！

'Yes. That's right. We call it *strangeness*, and denote it by s. The K^+ has $s=+1$; p^+, n^0, π^-, π^0, and π^+ have $s=0$; while Λ^0 and K^- have $s=-1$.

'Note that normal matter—protons and neutrons—has no strangeness. So to create a particle carrying

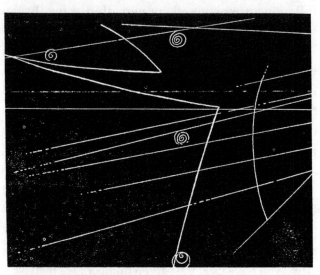

An associated production event

15 Visiting the 'Atom Smasher'

strangeness, you have to produce more than one of them at the same time: a particle with $s=+1$, and another with $s=-1$ (as in reactions (v) and (vii)). That way their combined s adds up to the original zero When the first examples of these new particles were found—before one got to know of s and its conservation—it was thought odd, or strange, the way they were always produced in association with each other; hence the name "strange". In fact, if I'm not mistaken, I think there's a photograph of an associated production event in your brochure. You might like to have a look at it. Anyway, since the discovery of strangeness, other properties have been identified: *charm*, *top* and *bottom*.

> 要引入的量子数越来越多了!

'So, what we find is that each particle involved in these collisions comes with a characteristic set of labels. For instance, the proton has electric charge, $Q=+1$, $B=+1$, $s=0$, and zero for its charm, top and bottom.

'But, you're doubtless thinking, this is all very well, but what has it got to do with finding out about the structure of the proton and neutron? After all, I did say earlier we could find out what protons are made of by looking at their close relatives—these new particles. It's at this stage we get involved in a piece of detective work. The basic idea is that we collect particles together that have certain properties in common: same B, same spin, etc. We then display them according to the values they have for two other properties. These are s, which we were talking about just now, and something else called *isotopic spin*, denoted by I_z. The name derives from the word 'isotope' meaning 'same form'. It arises from the fact that certain particles are so similar to each other—having the same strong interactions and almost identical masses—that one tends to think of them as different manifestations of the same particle. For example, the proton and the neutron are regarded as two forms of the same particle, the nucleon. In one of its forms, the nucleon has electric charge, $Q=+1$, in the other $Q=0$. In terms of isotopic spin,

> 又一个：同位旋！

they have values $I_z=+\frac{1}{2}$, and $I_z=-\frac{1}{2}$ respectively. (The 'spin' part of the name derives from the way it behaves mathematically in a way similar to ordinary spin.)

'One way of defining I_z is by the relation $I_z=Q-\overline{Q}$, where Q is the electric charge of the particle, and $\overline{Q}$ is the mean charge of the multiplet to which the particle belongs. So for example, with Q being +1 for the p, and 0 for the n, the mean charge for their nucleon doublet is $\overline{Q}=\frac{1}{2}(1+0)=\frac{1}{2}$. That in turn means I_z for the p is $I_z=1-\frac{1}{2}=+\frac{1}{2}$, and for the n, $I_z=0-\frac{1}{2}=-\frac{1}{2}$.'

'Right now, as I said, we take particles with certain properties in common and display them according to their individual values of s and of I_z. Take this one, for example...'

Dr Hanson sketched out an array of particles.

'This is one of the patterns we get: a grouping of eight baryons, each having $B=+1$ and spin $\frac{1}{2}$. Note the hexagonal shape, with two particles in the middle. You see it contains the neutron and proton. Set out like this, we begin to recognise that they are but two members of a family of eight.

> 粒子物理学家的分类法。

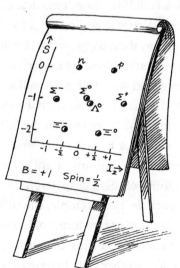

Sketched out an array Sketched out an array of particles

15 Visiting the 'Atom Smasher'

'Now look at this one...'

She drew a second pattern.

'This is the family of $B=0$, spin=0 mesons, containing the pions. It has exactly the same overall hexagonal pattern as before, consisting again of an octet, but this time with an additional singlet particle at the centre.

这里谈的相近的两幅图，不是很类似于当年门捷列夫搞的元素周期表吗？

'OK. What are we to make of this? Is it just a coinci-dence to get the same pattern repeated? No. To a mathe-matician, this pattern has a special significance. It deri-ves from a branch of mathematics known as "group theory"—a type of mathematics which, until recently, has had little application to physics beyond the description of the symmetries of crystals. We call this 'a representation of SU(3)'. 'SU' stands for Special Unitary, and describes the nature of the symmetry. The '3' refers to the three-fold symmetry. (Note how we get the same pattern when we rotate it through 120°, 240° and 360°.)

这种对称性就是物理学中美的表现之一。

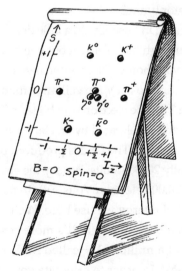

The same overall hexagonal pattern

'Besides coming up with this hexagonal octet pattern, the same SU(3) theory leads us to expect other patterns with three-fold symmetry. The simplest is a singlet. With the mesons, we have that as well as the octet. Then there is a decuplet making up a triangular pattern...'

At this point, Dr Hanson was interrupted by a knock on the door. She was handed a note.

'Oh good. Our minibus has arrived. I'm afraid I'll have to bring my mini-lecture to an abrupt close. Sorry about that, but I'm sure you will be covering this stuff on SU(3) representations later in your lectures.'

> 终于要走了，导游的这段开场白可够长的，在通常的游览中，听众肯定忍受不了导游这么啰嗦。

It was quite a long ride to their destination. On disembarking from the bus, they found themselves walking over to a very modest looking building.

'The accelerator's in *there*?' Mr Tompkins enquired of the guide, feeling somewhat let-down.

She laughed, and shook her head. 'No, no. It's down *there*!'

She pointed at the ground. 'About a hundred metres below the surface. This is just how we gain access to it.'

On entering the building, they took the lift. Emerging at the bottom, they found themselves at the entrance to the accelerator tunnel.

'Before going in I usually do a little demonstration at this point. You might not realise it but you have a particle accelerator in your home. There's one here for exa-mple,' she said, pointing to a TV surveillance monitor by the gateway. 'In a TV tube electrons are boiled off from a hot wire filament and accelerated by an electric field, so as to strike the front screen. The field is produced by a voltage drop of typically 20, 000 Volts. We say the electrons have an energy of 20,000 electron Volts(eV). In fact, the eV is the basic unit of energy we use here. Well, not exactly the eV; that's too small a unit. It's more convenient to deal in units of a million eV—called MeV, or 10^9 eV— denoted by GeV. In order to orientate you, the amount of locked-up energy in a proton is 938 MeV, or almost 1 GeV. I should perhaps also mention that we normally refer to the masses of particles in terms of their energy equivalents—rather than in terms of electron masses. So the mass of the proton is 938 MeV/c^2.

> 这里的"电子伏"（eV）可是能量单位，而不是电压单位。

15 Visiting the 'Atom Smasher'

'The particle accelerator you are about to see also accelerates electrons, but to much higher energies than in this monitor—energies sufficient to create those particles I was talking about. In fact we need to reach energies of tens or hundreds of GeV. That in turn requires the equiva-lent of a voltage drop of 10^{11} volts! But no way can we create and sustain such voltages—just think of the insulation problems. In a minute I'll be showing you how we get round this difficulty. But meanwhile, take a look at this...'

She reached in her pocket and pulled something out. She moved it over the face of the TV monitor. The picture immediately became grossly distorted.

'A magnet,' she said. 'Magnetic fields can be used for pushing particle beams around. That's another idea we shall be calling on. Incidentally, she added hastily, 'do not—I repeat, DO NOT—do this magnet experiment on

The tube through which the particles pass

> 要是电视转播这段，屏幕上就将在右上角打出"请勿模仿"的字样。

> 向下走100多米深，可得走上一阵子呢。

your TV set at home. It it's a colour set you'll wreck it; you will end up with a *permanent* record of what magnets can do to beams of electrons! It's only safe to do it with black-and-white sets like this one. OK. Let's move on.'

They walked down a passage which eventually opened out into a tunnel about the size of an underground railway tunnel. Opposite the passage opening was a long metal tube. It was 10 to 20 centimeters in cross-section and travelld the whole length of the tunnel. Going across to it, Dr Hanson explained.

'This is the tube through which the particles pass. They have a long way to go and mustn't hit anything, so it has to be evacuated. In fact, the vacuum in here is better than what you will find in many regions of outer space. This thing here,' she said, indicating a box wrapped round the tube, 'is a hollow copper radio frequency cavity. It generates the electric field responsible for accelerating the particles as they pass through it. But it's not particularly powerful—like the accelerating field in that TV monitor back there. So, how are we to reach the colossal energies we need?

'Well, take a look along the tunnel to the far end. Notice anything about the shape of the tunnel?'

They peered into the distance. Then a young man said, 'It's curved. Ever so slightly. I thought at first it was straight.But it's not?

'That's right. The tunnel—and the tube of the accelerator—is curved. It's actually circular; the whole th-ing is shaped like a hollow doughnut. The circum-ference of this, and other machines like it, is measured in tens of kilometres. What we are looking at here is but a tiny segment of the complete circle. The electrons are made to go round this circular racetrack. That means they eventually end up back where they started—all ready to go through the same radio frequency cavities again. Each time they pass through, they get an additional kick. At no stage do we need a big voltage drop.

15 Visiting the 'Atom Smasher'

Instead, we give them a series of kicks—small kicks—using the *same* cavities over and over again. Clever don't you think?'

They murmured their agreement.

'But this raises another problem. We have to bend the paths of the particles into a circle. How do you suggest we do that?'

'Well, based on what you did back there with the TV screen, I guess it must be done by magnets,' proposed Mr Tompkins.

汤普金斯先生真懂了不少物理知识。

'That's right. This is one here,' she said, moving on to a massive block of iron, again surrounding the tube. 'An electromagnet, with one pole above the tube and one beneath. That produces a vertical magnetic field, for bending the particles' path in the horizontal plane. Look along the tunnel, and you see there are lots of them, all the same, stretching right round the ring, producing the necessary circular path.

'A further problem is that the amount by which a magnet is able to deviate the path of a charged particle depends on the momentum of that particle—its mass times its velocity. But the particles are accelerating; they are constantly gaining in momentum. That means it becomes more and more difficult to bend their paths and hold them on course around the ring. So, what we have to do is this: As the particles gain in momentum, the electric current fed to the electromagnets is steadily increased. This in turn increases the strength of the magnetic field between its two poles. If the increase in magnetic field is synchronized so as to match the increase in particle momentum, then the particles stay on exactly the same course during their period of acceleration.'

exactly 说起来容易，做起来可不容易。

'Ah!' exclaimed the elderly gentleman. 'That must be why it's called a "synchrotron". I've always wondered.'

'Yes. That's right. It's a bit like hammer—throwing

at the Olympics—swinging a ball round and round in a circle, and having to hold on more tightly as it gets up speed.'

'So, do I take it these particles get released at some stage? You eventually let go, and they come out somewhere?'

'Well actually, no,' replied Dr Hanson. 'That's what we used to do. We would activate a kicker magnet or electric field to eject partl; cles from the accelerator once they had reached maximum energy. They then hit targets of copper or tungsten where the new particles were produced, and these were then sorted and separated out by more magnetic and electric fields. Eventually they were led off to detectors, like bubble chambers.

'The trouble with fixed targets like that was that as far as useable energy was concerned, it was not very efficient. You see, in a collision one must conserve not only energy but also momentum, or impetus. A projectile from an accelerator carries momentum, and this must be passed on to the particles emerging from the collision. But the final particles cannot have momentum without also having kinetic energy. So, in effect, some of the projectile's energy had to be held back in reserve— so that it could later be passed on to the final particles as the kinetic energy to go along with the necessary momentum.

> 注意，作者用 beauty 来形容这台大机器。

'The beauty of this machine is that it has two beams going in *opposite* directions. With a head-on collision, the momentum brought in by one particle is balance by an equal and opposite momentum brought in by the particle moving in the opposite direction. That way, *all* the energy, brought in by both beams, becomes available for particle production. It's a bit like a head-on collision between cars; this is far more devastating than a collision where one of the cars was station ary, and they just shunt along afterwards.'

> 是啊。想想两辆车迎头对撞时的情形吧，肯定能撞出一些汽车的组成部件来。

'So, does that mean you have two accelerators— one for each beam?' Maud asked.

15 Visiting the 'Atom Smasher'

'No, that's not necessary. A particle with negative electric charge gets bent by a magnetic field in the opposite sense to that of a positively charged particle. So what we do is send positive particles round one way, and negative ones the other way—using the same bending magnets and accelerating cavities. Of course, to stay on exactly the same track, they must at all times have the same momentum, so the two sets of particles must have the same mass as well as the same speed. That's why in here we have counter-rotating electrons and positrons. Another combination would be protons and antiprotons.

'So, the particles are accelerated round and round in opposite directions until they reach maximum energy. They are then brought together at selected points around the ring so as to collide head-on. And it is at those intersection points we set up our detection devices.'

'This head-on business seems the obvious way of doing thingsfor the reason you stated. So why did they ever bother with fixed targets in the first place?' enquired the elderly man.

'The difficulty with these colliding beams is that of getting an intense enough beam of positrons or of antiprotons,' the guide explained. 'We concentrate them into tight bunches about the size of a pencil. But even so, when the beams are brought together, most particles go sailing through the intersection point without encountering any particle from the other beam. Highly sophisticated techniques have to be used for con centrating the particles so as to give a worthwhile number of collisions. That's done by the focussing magnets, like this one here,' she said, pointing out a different type of magnet. 'This has two pairs of poles instead of the usual single pair.'

'But what I don't understand is why the machine has to be so big,' asked a woman.

'Well, youhave to realise that there is a certain

> 机器造得越大，花钱也就越多。

maximum field one of these magnets can produce. (As particle energies go up, they are more difficult to steer, so to close the circle, you must have more and more of these magnets. But, as you see each magnet has a certain physical size—about six metres. So, given the number of magnets that have to be fitted in round the circle—about 4000—to say nothing of the focussing magnets and the accelerating cavities, that sets the size of the circle. The higher the final energy of the particles, the bigger the circle has to be.'

'And are the particles going round in there right now?' asked one of the party.

'Oh goodness me, no!' exclaimed Dr Hanson. 'When the machine is operating no-one is allowed down here in the accelerator tunnel—the level of radiation wo-uld be too high. No. This is one of the periodic, routine shut-down periods for maintenance. That's why your visit was timed for today.'

Taking a quick look at her watch, she continued. 'Right, we must be moving along. Follow me, please, and I'll take you to one of the points where the beams collide. It'll give us a chance to take a look at one of the detectors.'

Having walked a very considerable distance past the seemingly endless succession of magnets, they eventually reached a section where the tunnel opened out to become a vast underground cavern. There in the centre, and towering above them, was an object as big as a two-storey house.

'That's the detector,' Dr Hanson announced. 'What do you think of *that*?'

They were suitably impressed.

> 花这么多钱造加速器，时间就是金钱啊。若不是要检修，停机游览的费用普通人肯定承受不了。

'No. Please don't go wandering off,' she hastily called to a couple who were making there way over to take a closer look, 'We mustn't get in the way of the physicists and technicians. They are working to a very tight schedule. All their maintenance has to be carried out

15 Visiting the 'Atom Smasher'

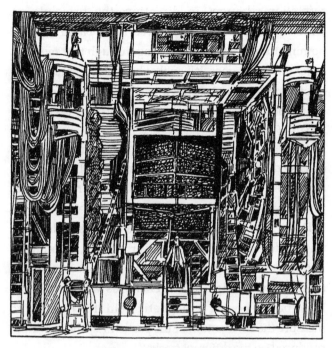

An object as big as a two-storey house

during this brief shut-down period.'

She went on to explain how the detector was wrapped around the tube at one of the intersection points of the beams. Its purpose was to detect the particles coming out of the collisions. In fact, it was not one detector, but many, each with its own characteristics and its own job to do. For example, there were transparent plastics that scin-tillated when charged particles passed through them. There were materials where the particles emitted a special kind of light (called Cerenkov radiation) whenever a particle passed through at a speed greater than that of light in that medium.

'But I thought relativity theory said nothing could travel faster than light—the ultimate speed barrier,' a woman interrupted.

'Yes that's true—but only if one is thinking of the speed of light in a *vacuum*,' Dr Hanson explained. 'When light passes into a medium such as water, glass, or plastic, it slows down. That's the reason you get refraction—changes in direction—the principle on which the spectacles you are wearing is based. There's nothing to stop a particle passing through that medium faster than light does. When that happens, it emits a kind of electr—omagnetic shock wave—analogous to the sonic boom given out by aircraft when they exceed the speed of sound.'

> 这些讲述太技术性了，不知伽莫夫本人亲自操刀情形会如何。

She went on to describe how some detectors consisted of gasfilled chambers containing thousands of electrified fine wires. When a charged particle passes through the chamber, it knocks electrons off the atoms belonging to the gas (it ionises them). These electrons migrate to the wires where their arrival can be recorded. In this way the track of the particle can be reconstructed from a knowledge of which wires were affected. By superim-posing a magnetic field it is possible to measure the momentum of the particles by the curvature produced on the different tracks.

Then there were calorimeters. These are so-called after the calorimeters often used in school science lessons for heat experiments aimed at measuring energy. The calorimeters used here measure the energy of individual particles, or the total energy of close bunches of particles.

Knowing the energy of a particle, and combining it with a knowledge of the momentum derived from the magnetic curvature of the tracks, one can identify the mass of the particle coming from the primary interaction. Finally, outside the calorimeters are chambers which act as detectors of muons. The muon is a particle which, like the electron, does not experience the strong nuclear force. But unlike the electron, it does not easily lose energy through emitting electromagnetic radiation (on account of it being some 200 times heavier than the electron). It can therefore power its way through most obstacles without doing much. And that is the very property used against it. The outside muon detector is packed with dense material.

15 Visiting the 'Atom Smasher'

Anything getting through that has to be a muon!

All these different types of detector are arranged like the layers of a cylindrical onion wrapped around the segment of the accelerator's tube where the interactions take place. They have to be fitted together rather like a gigantic three-dimensional jigsaw puzzle. In total the structure weighs 2,000 tons.

'But all this happens only when the synchrotron is switched on, surely,' said Mr Tompkins.

'Of course.'

'But no-one is allowed down here when it's on. So how do the scientists know what is happening in there?'

'Good point,' remarked Dr Hanson. 'See all those?' she said, pointing to a tangled web of cables leading away from the detector. To Mr Tompkins it looked like a spaghetti factory hit by a bomb.

'They take the electronic signals from the individual detectors and pass them on to the computer. The computer processes all the information and reconstructs the tracks of the particles. These can then be displayed to the physicists in the remote control room. That over there shows the sort of thing they have to deal with.'

She nodded towards a photograph sellotaped to the wall.

'Come and have a quick look at it. Then I'll take you to see the control room itself.'

Following the others, Mr Tompkins momentarily glanced back at the detector. In doing so, he failed to notice that one of the maintenance technicians had left a cable running across the floor. He tripped over it, fell, and hit his head on the concrete floor...

唉，汤普金斯先生终于出问题了，接下来该容易点儿了吧。

'Good heavens, Watson, this is no time to take a rest. Get up man and give me a hand.'

A figure dressed as Sherlock Holmes was standing over him. Mr Tompkins was about to explain that his name was not Watson, when his attention was distracted by the detector. It was spewing out particles in all

科学家的工作很像侦探福尔摩斯的工作，不过福尔摩斯要真看到这些，准晕。

227

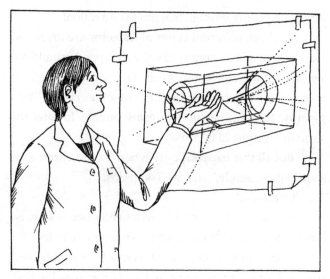
Displayed to the physicists in the remote control room

directions! They were rolling about all over the floor.

'Come on! Collect them up for me—as many as you can carry,'

Mr Tompkins looked around for Dr Hanson and the rest of the party. They were nowhere to be seen. He concluded they must have gone off to the control room without him. Odd that, but presumably they would come back for him at some stage. Meanwhile he thought he had better humour this madman in fancy dress.

Gathering up an armful of the particles, he carried them over to the Holmes character, who was quietly surveying neat arrangements of particles laid out on the floor. Mr Tompkins recognised them as the familiar hexa-gonal shapes of the SU(3) representations.

'Right. So much for spin $\frac{1}{2}$. Now for the spin $\frac{3}{2}$, $B=1$ particles,' said Holmes holding out his hand.

'I beg your pardon.'

'Particles with spin $\frac{3}{2}$ and $B=1$. Come along my dear fellow; I've done the others.'

Mr Tompkins was confused. 'How am I to know...'

'Look at the labels,' said the great detective wearily. It was only now that Mr Tompkins noticed that each particle had a tiny label stuck to it. It listed the particle's

15 Visiting the 'Atom Smasher'

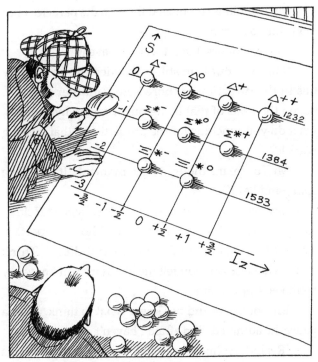

Particles with spin $\frac{3}{2}$ and $B=1$

properties. Sifting through them, he handed over those stating that they had spin $\frac{3}{2}$ and B=1. Holmes bent down and laid them out on the floor. After somere arranging, he drew up a chair, and sat studying them.

'Well, Watson,' he murmured. 'What do you make of it? Let me hear your reconstruction of the situation.'

Mr Tompkins gazed at the pattern before him.

'It looks like a triangle,' he ventured.

'You are of that opinion, are you? As a man of precisely scientific mind, do you recognise anything erroneous with that conclusion?'

'Well, the bottom apex is missing.'

'Exactly! As you astutely observe the triangle is incomplete. A particle is missing. May I have the final piece?'

Still surveying the pattern, Holmes again held out his hand.

Mr Tompkins rummaged through the particles once more, but to no avail.

'Sorry, Holmes. I don't seem to have it.'

'Hmmm. But I am still persuaded the probability lies in the direction of there being another particle. So, on the basis of that being our working hypothesis, what do you conclude for the properties of that missing particle?'

Mr Tompkins thought for a moment. 'It will have spin $\frac{3}{2}$ and $B=1$?'

'My dear Watson, you excel yourself,' sighed Holmes sarcastically, 'Of course, it will have those properties; it would not belong to this family if it didn't. Think man! What else can you tell me about it? You know my methods. Apply them!'

Mr Tompkins did not know what to think. After a pause, he admitted, 'I'm afraid I haven't a clue.'

'Really!' Holmes exploded. 'It is perfectly obvious to the trained man of science that the missing particle is negatively charged, having no positive or neutral counterpart—a very singular particle; it has s=−3 (quite an unprecedented amount of strangeness incidentally); and a mass of about 1680 MeV/c^2.'

> 汤普金斯这时真把自己当成华生医生了,可是华生的原型并不是很聪明啊。

'Good Heavens, Holmes, you astound me!' exclaimed Mr Tompkins. With a start he realised that, unwittingly, he had now completely assumed the role of Watson.

'As it is the last particle to complete this pattern, I shall call it the Omega minus (Ω^-),' concluded Holmes.

'But I don't understand. How do you know all this?'

The great man smiled. 'It is a pleasure for me to exercise any small powers which I possess at your expense. In the first place how many gaps are there in the pattern?'

'One.'

'Precisely. Hence we are dealing with a single missing particle. Next, what do you note about its strangeness?'

15 Visiting the 'Atom Smasher'

'Well, the gap in the pattern is level with $s=-3$.'

'Exactly. And now its electric charge?'

'No. I'm afraid I'm stumped on that one.'

'Use your powers of observation, man. What do you notice about the electric charge of the particles at the extreme left hand end of each row?'

'They are all negative.'

'Quite. Our Ω particle is at the left hand end of its row, so it too must be negative.'

'But,' protested Mr Tompkins, ' it's the only particle in that row. So it is also at the right hand end of the row as well.'

'What of it? Run your eye down the right hand members of each row. What do you notice?'

Mr Tompkins studied them for a moment and then declared, 'Oh I see, what you mean. Each successive row loses a unit of charge: $Q=+2, +1, 0$, so the last one will be $Q=-1$, which is what we had before. But then you said something about the mass of the Ω. How could you possibly decide that?'

'Examine the masses of the other particles.'

'Yes. So what?' asked Mr Tompkins thoroughly perplexed.

'Mental arithmetic! What are the differences in mass between successive rows of particles?'

'Er. Between the Δ s and the Σ *s, I make that 152 MeV/c^2. And between the Σ *s and the Ξ *s ... 149 MeV/c^2. They're almost the same.'

'From which I surmise the same difference might apply between the Ξ *s and our conjectured Ω particle. Our net is closing in. With those properties in mind, perhaps you would be so kind as to go and look for it.'

With that, Holmes leant back in his chair, placed his finger tips together, and closed his eyes.

Irritated though Mr Tompkins was with Holmes' condescension, he was intrigued to know whether there was any truth in these deductions. So he dutifully trudged

福尔摩斯还是那么玄，甚至更玄了。

off with the intention of rummaging through the various particles strewn over the floor around the detector.

But before he got there, a jostling crowd of electrons appeared from nowhere. Mr Tompkins found himself completely surrounded and caught up with them.

'All aboard!' came a commanding voice. Immediately all the electrons surged towards the accelerator, propelling Mr Tompkins with them. They crammed inside the tube; it was worse than catching a rush-hour train. Everyone was furiously using their elbows to repel the others and make space for themselves.

'Excuse me. But what's happening?' Mr Tompkins enquired of a neighbouring electron.

'What's happening? You new around here or something?'

'As a matter of fact...'

'Then welcome to the kamikazes!' the electron leered menacingly.

'Sorry? I don't ...'

But there was no time for explanations. A violent shove in the back, and they were on their way, heading down the tube. Just as Mr Tompkins thought he was bound to crash into the curving wall, he became aware of a steady sideways force directing him away from it.

'Ah!' he thought, 'that must be the effect of the bending magnets.' Another shove in the back. 'And that must be another accelerating cavity we've gone through.'

As they coasted along between the periodic kicks, he noticed that the bunch of electrons tended to disperse somewhat. 'I suppose that's because we all have negative electric charge and are repelling each other.'

But then they were abruptly squashed together again. He surmised that must have been due to them passing through a focussing magnet.

Suddenly, to Mr Tompkins alarm, out of the gloom

15 Visiting the 'Atom Smasher'

a swarm of particles came flying towards them from the opposite direction. They narrowly missed.

'Help!' Mr Tompkins cried. He turned to his companion. 'Did you see that? I mean, that was DANGEROUS. Who were *they*?'

'You *are* new, aren't you,' came the sneering reply. 'Positrons! Who else?'

The pattern of events repeated itself: a succession of accelerating kicks, interspersed with focussing episodes, and all the time the bending magnetic field getting stronger and stronger the more energetic the particles became. And of course, at regular intervals, there were the bunches of positrons flying past doing the circuit in the opposite sense.

In fact, things were beginning to get very ugly. Now every time the positrons went past they shouted abuse. 'Just you wait. We're going to get you lot!' they taunted.

'Oh yeah. You and who else?' retaliated Mr Tompkins's electrons. Both electrons and positrons seemed gripped by a mounting sense of anticipation and excitement.

But Mr Tompkins was ceasing to care. With each turn of the accelerator he felt more and more dizzy and nauseous. Until, that is, his attention was suddenly arrested by a muttered warning from his companion: 'OK. Brace yourself. Full energy. This is it. Good luck! You'll need it.'

Mr Tompkins was about to ask what he meant by this-but there was no need: the positrons were upon them. But now they were heading straight for each other. On all sides, Mr Tompkins saw violent collisions between electrons and positrons, each producing jets of new particles spreading out in all directions. No sooner had some of the new particles been created in the collision, than they split up into other particles. Eventually all the debris passed through the walls of the accelerator tube and disappeared from sight.

Silence. It was all over. The positrons had gone,

> 想想开车逆行的感觉吧。

> 就这么转，能不晕吗！

leaving just the electrons. On looking a round him, it seemed to Mr Tompkins that despite all the violence that had taken place, most of the electrons, like himself, were completely unscathed.

'Phew! That was lucky, the sighed with relief.' I'm glad that's over.'

His companion shot him a scornful look. 'It's really amazing,' the electron remarked. 'You honestly don't know anything *at all*, do you.'

With that the positrons were back! The whole frightening sequence was reenacted a second time, then a third time, a fourth time, and so on: periods of quiet, inter-spersed by mayhem. Mr Tompkins gradually came to realise that the collisions always took place at the same points around the ring. 'These must be where the detectors are placed,' he guessed.

> 汤普金斯先生自己要是真的撞上了，会摔成什么样？

It was during one of the encounters between the beams, that it happened—what Mr Tompkins had most dreaded. A direct hit? Without warning, he was knocked flying. He was sent clean through the wall of the accelera-tor where—as he had suspected—the detector was awai-ting him. He was only dimly aware of what occurred next:strong bending to one side, showers of sparks, flaslles of light, and a series of bumps as he crashed through metal plates, before eventually coming to rest in one of them. How he managed to extract himself from the plate he could never remember; he was much too dazed. But somehow he did it, and found himself once more in the experimental hall, amid a pile of other particles that had also leaked out of the detector.

He was lying there staring up at the ceiling, trying to collect his wits, when a voice coyly asked, 'Looking tor me?'

At first he did not realise the question was being directed at him. But when the seductive enquiry was repeated, he struggled to a sitting position.

'I beg your pardon,' he ventured, looking around.

15 Visiting the 'Atom Smasher'

He found he was being addressed by one of the particles—a rather exceptional, indeed exotic-looking particle.

'I don't think so,' he mumbled.

'Are you sure?' she persisted.

'Quite sure.'

There was an awkward pause. 'Pity. I could do with company—being single. You might at least look at my labels,' she added sulkily.

Mr Tompkins sighed, but dutifully complied. He read out 'Spin $\frac{3}{2}$, $B=1$, negative charge, $s=-3$, mass 1672 MeV/c^2.'

'So?' she said expectantly.

'So what?' he replied, wondering what she was getting at. But then it struck him. 'Good Lord! You're...You're the Ω^- particle! You're the one I've been sent to look for! I'd quite forgotten. Oh my goodness! I've fo-und the missing Ω^-!'

Excitedly gathering her up, he rushed back to Holmes to show him his prize.

又变回去了。

'Excellent!' exclaimed Holmes. 'Just as I thought. Put it where it belongs.'

Mr Tompkins placed it on the floor to complete the triangular decuplet. Holmes lit up his favourite black clay pipe, and sat back puffing contentedly.

'Elementary, my dear Watson,' he declared. 'Elementary.'

Mr Tompkins gazed for a while at the patterns laid out before them, the hexagonal octets and the triangular decuplet, but then became aware of just how acrid the fumes from Holmes' strong tobacco were. He was becom-ing enveloped in smoke. It was most unpleasant, so he decided to move away.

Wandering off, he idly decided to do a tour right round the detector. Reaching the far side, he was surprised and delighted to see a familiar figure hunched over a work bench. It was the woodcarver!

> 又见到木雕匠了，不过这次他是在给夸克涂色了。

> "华生"也有资本看不起福尔摩斯了。

'What are you doing here?' he enquired.

The woodcarver looked up. On recognising his visitor, his face broke into a smile. 'Well, if it isn't you! Good to see you again.'

They shook hands.

'Still busy with your painting, I see,' commented Mr Tompkins.

'Ah, but I've moved on since the last time you saw me,' he said. 'New job. No more painting protons and neutrons. These days it's *quarks*!'

'Quarks!' exclaimed Mr Tomkins.

'That's right. The ultimate constituents of nuclear matter. It is what protons and neutrons are made of.'

He looked about him, and beckoned his friend to come closer. 'Couldn't help overhearing your conversation with that loudmouth over there,' he muttered confidentially.' "Elementary, my dear Watson. Elementary" 'he repeated mockingly.' Take it from me, he doesn't know what he's talking about. Elementary, my foot! Those particles of his are anything but elementary. Take my word for it: quarks is what it's all about.'

'So, what exactly are you doing?' asked Mr Tompkins.

'Painting the quarks,' the woodcarver replied. 'As the new particles come out of the accelerator, I paint their quarks.'

Taking up a finely pointed brush in one hand, and a pair of tweezers in the other he continued. 'Fiddly work. The quarks are so small. Look. Here's a meson. See the quarks in there: one quark, and an antiquark. I get hold of the quark like this, ' he said, reaching inside with the tweezers and catching hold of the quark. 'You can never pull the quarks out; they're glued in too tightly. But no matter. I can paint them perfectly well while they're still inside. I paint the quark red, like this. Then. with this other brush, the antiquark cyan.'

'Those were the colours you used for the proton and

15 Visiting the 'Atom Smasher'

electron,' recalled Mr Tompkins.

'That's right. And as you see the combination makes the overall meson white. But I can also use these other complementary colour combinations: blue with yellow, and green with magenta(or purple),' he said pointing to other paint bottles on the bench.

'The baryons, like this proton here, are made from a combination of three quarks. For baryons, I paint each quark a different primary colour: red, blue, and green. That is the al ternative way of producing white; you either use a colour and its complementary colour, or you have a mix of all three primary colours.'

Mr Tompkins's thoughts wandered to his earlier encounter with the monk. He imagined Father Pauli would approve of mesons—the marriage of opposites, but was not sure what he would have made of the combination of three of the same!

The woodcarver continued gravely, 'I would have you know this is vitally important work. The very fabric of the Universe depends on what I am doing here. Painting protons and electrons was just to make them look pretty—more easily distinguishable in diagrams in popular physics books. This, on the other hand, is serious colour. I mean, it's something *physicists* themselves refer to. It explains why quarks hold on to each other—why they can't come out separately. To be able to stand on its own, a particle has to be white-like the protons and neutrons I've just finished doing-over there in that box. They are now ready for shipment. Individual quarks, on the other hand, are coloured, so they must for ever be attached to others of the appropriate colours. I trust I have made this all clear.'

> 好大的口气，简直就像上帝。

Mr Tompkins felt that some of the information he had previously read in the brochure might now be slotting into place. But why exactly particles had to be white was still a mystery to him. He went over to the box of nucleons and lifted the lid. He was struck by their

brilliant whiteness. In fact he was quite dazzled by it, and had to shield his eyes...

'I do believe he's coming round at last.' It was the voice of Maud. 'The light. Please! You're blinding him. Darling, darling, are you all right? What a relief! We were so worried. What a knock you took! How are you feeling?'

'It was the positron,' Mr Tompkins murmured. 'The positron hit me.'

'A positron hit him?!' a voice enquired. 'Is that what he said?'

'Concussion,' pronounced another. 'He's suffering from concussion. Rambling. We must get him over to Casualty right away. He'll need to take it easy for a while. And we must get a dressing on that cut to his forehead.'

虽然醒过来了，可还晕着呢。

16 The Professor's Last Lecture

教授的最后一篇演讲

Ladies and gentlemen:

It was Murray Gell-Mann and Yuval Ne'eman who in 1962, independently of each other, recognised that the particles could be gathered together into family patterns based on the SU(3) group.

Not all the patterns were found to be complete; there were gaps. In this respect, the situation bore similarities to that which had earlier confronted Mendeleéff when compiling his Periodic Table of the atomic elements. He too had found recurring patterns of behaviour—provided he left gaps for hitherto undiscovered elements. By looking at the properties of the elements adjacent to the gaps, Mendeleéff had been able to predict the existence and nature of the unknown elements. History now repeated itself as Gell-Mann and Ne'eman predicted the existence, and the detailed properties, of the Ω^- on the basis of a gap in a triangular decuplet. The remarkable discovery of the Ω^- in 1963 served to convince the scientific community of the validity of the symmetry group SU(3).

The Mendeleéff table, through setting out the relationships between the atomic elements, hinted at an inner composition; the elements were to be regarded as different variations on a common theme. This suggestion was later to be borne out in the theory of atomic structure, whereby all atoms consist of a central nucleus with surrounding electrons.

In 1964, Gell-Mann and George Zweig suggested

> 由伽莫夫的合作者写的这篇教授的最后演讲，仍然不那么好懂。也许教授真的太老了，也许，就只能归结于写作者在普及性文字上的功力与伽莫夫相比还有一些距离吧。

that the similarities and family patterns displayed by the particles were likewise a reflection of some inner structure. This proposal held out the possibility that the 200 or more particles, until that time regarded as 'elementary,' were in fact composites constructed from yet deeper fundamental constituents. These constituents were to be called *quarks*. At the present time, we believe quarks to be truly elementary. They are treated as point-like, having no inner structure consisting of 'subquark' constituents. But who knows? We might be proved wrong-again!

> 夸克的名字，最初还是从乔伊斯的小说中借用过来的。

The original scheme was based on there being three types, or *flavours*, of quark: up, down, and strange quarks. The first two were so named because of the 'up' and 'down' direction of their isospin. The strange quark got its name through carrying the newly discovered property of matter: strangeness. The recognition in the 1970s of particles carrying two further properties(charm and bottom), and in the 1990s, yet another property(top), was later to necessitate the inclusion of three further flavours of quark-those carrying the additional properties. The properties of all six quarks are set out in Table 1.

In addition to these six quarks, there are six antiquarks possessing the opposite values of all the quantities displayed in the table. For example, the $\bar{s}$ antiquark to the s quark, has $Q=+\frac{1}{3}$, $B=-\frac{1}{3}$, and $s=+1$.

> 真够复杂的。

Table 1 The properties of the quarks

	Q	B	s	c	b	t
d	$-\frac{1}{3}$	$\frac{1}{3}$	0	0	0	0
u	$\frac{2}{3}$	$\frac{1}{3}$	0	0	0	0
s	$-\frac{1}{3}$	$\frac{1}{3}$	−1	0	0	0
c	$\frac{2}{3}$	$\frac{1}{3}$	0	0	0	0
b	$-\frac{1}{3}$	$\frac{1}{3}$	0	0	−1	0
t	$\frac{2}{3}$	$\frac{1}{3}$	0	0	0	1

Q is the electric charge; B is the baryon number; s the strangeness number; c is charm; b is bottom; and t is top; d, u, s, c, b, and t denote the six quarks.

16 The Professor's Last Lecture

From these quarks and antiquarks, all the new particles produced in high energy collisions can be synthesised. The baryons are made up of three quarks: (q, q, q). So, for instance, the proton is the combination (u, u, d), the neutron is (u, d, d), and the Λ^0 is (u,d,s). You should check from the table that these combinations do indeed yield the properties of the particles (the proton, for example, having $B=+1$, and $Q=+1$).

The antibaryons consist of three antiquarks: $(\bar{q}, \bar{q}, \bar{q})$. This results in baryon and antibaryon having exactly opposite properties.

What of mesons, such as the pion? Mesons are constructed from the combination of a quark and an antiquark: $(q, \bar{q})$. Thus, for instance, the π^+ consists of $(u, \bar{d})$. Again check that this combination yields the right overall properties for that pion: $B=0$, and $Q=+1$.

I should point out that not *all* particles are made of quarks. Only the baryons and mesons are so constructed. In fact, we call all such particles collectively, *hadrons*—a word meaning 'strong'. Hadrons feel the strong nuclear force; other types of particles, such as the electron, muon and the neutrinos, do not. They are collectively known as *leptons*. Indeed, the names 'baryon' and 'meson' can be something of a misnomer. They are based on the idea of how massive the particle is. But we now know of a lepton, the tau, which is twice as heavy as the proton—hardly a 'light' particle! It is therefore preferable to delineate particles according to whether they are hadrons (strongly interacting), or leptons (not subject to the strong nuclear force).

照此说，夸克又不是万物之源了。

So far we have talked of quarks bound up in hadrons. What of free quarks? They should be easily recognisable with their fractional onethird or two-thirds electric charge.

Despite strenuous efforts, none has ever been seen. Even in the highest energy collisions, quarks are never ejected. This calls for an explanation.

One idea canvassed for a time was that quarks were not real ; they were mere mathematical entities-useful fictions. The particles behaved *as if* they consisted of quarks, but there was no such thing as an actual quark.

But then came a conclusive demonstration of their reality. It was a further example of history repeating itself. Recall how Lord Rutherford in 1911 had demonstrated the existence of the nucleus by firing projectiles(alpha particles) at atoms, and observing some rebound at large angles. This indicated that the projectiles had struck a small concentrated target (the nucleus) within the atom. In 1968, it became possible to fire high energy electrons into the *interior* of the proton. Evidence began accumulating that the electrons were occasionally suffering large sideways kicks, indicative of them having rebounded from some small and concentrated electric charge inside the proton. This was confirmation that the quarks were indeed real. Indeed, from the frequency of the large-angle scatters, it could be calculated that there were three quarks inside the proton.

> 注意，这里讲的，还是间接地"计算"出，还不是直接观察。

So, if the quarks are definitely in there, why do they never come out singly? Also, we need to address the question as to why we get only (q,q̄)and (q,q,q) combinations. Why not others, such as (q,q̄,q̄) and (q,q,q,q)? To elucidate this we turn to the nature of the force between the quarks.

We begin by recalling how the attraction between the proton and electron of a hydrogen atom arises out of the electrostatic force operating between the electric charges carried by the proton and electron. By analogy therefore, we are led to introduce an additional kind of 'charge'. We postulate that quarks carry this kind of 'charge' (in addition to electric charge), and the strong force arises because of interactions occurring between these 'charges'. For reasons that will become clear later, we call it *colour charge*.

In the same way as opposite electric charges attract

each other, so opposite colour charges attract—only with a much stronger action. We postulate that quarks carry positive colour charge, and antiquarks negative. This accounts for the ready occurrence of the $(q,\bar{q})$ meson combination. Again in analogy with the electrostatic case, we assume that like colour charges repel each other. This accounts for the nonexistence of $(q,\bar{q},\bar{q})$. Just as a second electron close to a hydrogen atom does not attach itself because its attraction to the proton is compensated by its repulsion for the electron already there, so a second quark does not attach itself to a meson because of the repulsion of the other quark.

But what, you will be asking, is the explanation for the (q,q,q) combination? It is here we must take note of a *difference* between electric charge and colour charge. Whereas there is only one kind of electric charge (which can be positive or negative), there are *three* kinds of colour charge (each of which can be positive or negative). We call them red, green, and blue—or r, g, and b—for reasons that will become clear soon. (I immediately hasten to emphasise, however, that they have nothing to do with ordinary colours.) There being three types of colour charge, this prompts the question: What kind of interac-tion takes place between quarks carrying different types of colour charge—for instance, a q_r carrying red, and a q_b carrying blue? The answer is that they attract each other. This force of attraction is such that the combination (q_r, q_g, q_b), where the three quarks are each of a different colour—so each is attracted to the other two—is parti-cularly strongly bound and stable; hence the occurrence of baryons.

Why do we not get the (q,q,q,q) combination? Because, given that there are only three types of colour charge, the fourth quark must have the same colour charge as one or other of the three quarks already in the baryon. It will therefore be repelled by its like charge. It turns out that this repulsive force exactly cancels out the attraction the fourth quark experiences from the other two quarks of different colour charge. Hence it does not attach itself.

At this point we can begin to understand why the name 'colour' charge has its appeal. Just as atoms are normally electrically neutral overall, we say that the allowed combinations of quarks have to be colour neutral—or 'white'. There are two ways of mixing colours to produce white. Either a colour is combined with its complementary (or negative) colour, or the three primary colours are combined. But those are exactly the rules for combining the colour charges to produce the overall neutral combinations: the meson and the baryon.

So summarising, quarks carry a positive amount of eitherr, b, or g,while an antiquark will have a negative, or complementary amount: $\bar{r}$, $\bar{b}$, or $\bar{g}$. Like charges repel, so for example r repels $\bar{r}$: g repels $\bar{g}$. Opposite charges attract, hence r attracts r ,etc. Finally, charges of different type attract each other.

A question we have yet to address is that concerning the absence of isolated quarks. To answer this we must understand more deeply the nature of the colour force, and indeed of forces in general.

In the spirit of quantum physics, where interactions occur discretely rather than continuously, we regard the mechanism by which a force—*any* force—is transmitted from one particle to another as involving the exchange of an mtermediary third particle. Basically we can think of particle 1 emitting the intermediary in the direction of particle 2, and in the process suffering a recoil-much as a rifle recoils in the opposite direction to the motion of the bullet. Particle 2, on receiving the intermediary, takes up its momentum, causing it to recoil away from particle 1. The overall effect of this exchange is that both particles are pushed apart. The process then repeats itself when the intermediary is returned; there is a further pushing apart. The net effect is that the two particles repel each other; i.e. they experience a

> colour charge, 色荷。只是其中的"色"字又与我们通常意义上的颜色意义大为不同。

> 在这种理论中，对于两个粒子间的相互作用的产生，存在有第三个参与"交换"过程的中介粒子是关键。

repulsive force.

What about forces of attraction? Essentially the same mechanism is involved, though I suppose this time—if you insist on having an analogy—we must think of the particles throwing boomerangs rather than shooting bullets! Particle 1 emits the intermediary in a direction *away* from particle 2, hence experiencing a recoil *towards* that particle; the latter then receives the intermediary from the opposite direction, and is also pushed towards its companion.

In the case of the electric force between two charges, the intermediary particle is the photon. The two charges are either repelled or attracted due to the repeated exchange of photons.

That being so, it prompts us to ask whether the strong force between the quarks is also open to an explanation in terms of the exchange of some kind of intermediary particle. The answer is yes; quarks are held together in the hadron by the exchange of particles called *gluons*. (I take it I need not explain the origin of *that* name!) There are eight different types of gluon. These arise because, in the process of exchanging a gluon, the quarks retain their fractional electric charge and fractional baryon number, but are able to exchange their colour charge. The gluon, on being emitted by the first quark, carries away the original colour charge of that quark. But the quark cannot be left colourless, so at the same time as it loses its original colour, it is invested with the colour of a second quark. The gluon, on arriving at the second quark, cancels out this quark's original colour charge, while transferring to it the colour charge taken from the first quark. The net result is that the quarks have swapped colour charge.

For these transformations to occur, the gluon must possess both a colour charge and a complementary colour charge. For instance the gluon, $G_{r\bar{b}}$, will have charges r and. $\bar{b}$. It can take part in the following transformations:

$u_r \to u_b + G_{rb}$ followed by $G_{rb} + d_b \to d_r$

With there being three colour charges and three complementary colours, that makes for 3 × 3=9 different possible combinations of a colour and a complementary colour. These split up into an octet and a singlet. (Recall the octet and singlet we came across when assigning the mesons to SU(3) representations.) The singlet state for the gluons would correspond to a mixture of $r\bar{r}$, $b\bar{b}$ and $g\bar{g}$. Being colour-neutral, it would not interact with the quarks, and for that reason we neglect it. Which leaves the octet, i.e. eight gluons in all.

Gluons, like photons, are massless. But unlike photons, which are not themselves electrically charged, gluons—as we have just noted—do carry colour charge. They therefore not only interact with quarks, but also with *themselves*. This dramatically changes the character of the force they transmit. Whereas the electric force gets weaker the further apart the electric charges (falling off as the inverse of the square of the distance between them), the colour force has the same value regardless of the separation (apart from when the colour charges are very close, when the force becomes almost non-existent—rather like an elastic band becoming slack when its ends are close). Thus, when quarks are close together, there is very little force between them. But increase the separation, and the force attains a constant value.

With this in mind, let us return to the question of why no isolated quarks are found. Suppose we were to try and separate two quarks. Because of the constant force between them, it takes more and more energy to increase their separation. Eventually you reach the point where you have put so much energy into stretching the bond binding the two quarks that there is enough there to create a quark—antiquark pair. And that's what happens: the bond snaps and a pair is created. The antiquark of the new pair goes off with the ejected quark to form a meson,

> 曾有科学史家研究过，为什么这些形形色色的"子"的名称大多以"-on"结尾。

> 这可是个重要的、大家关心的问题。

16 The Professor's Last Lecture

while the quark of the new pair is left behind in the hadron to take the place of the old quark. The situation is rather similar to what happens when one takes a bar magnet and tries to isolate the north pole from the south pole. Snapping the magnet in half, new north and south poles are generated, leaving us with two bar magnets; we are no closer to the goal of having an isolated pole. In similar vein, breaking the bond between the quarks does not result in an isolated quark.

> 可是，磁铁虽然两极不可分，却与人们习惯的要找的最基本的物质基元有所不同。

We have talked about the proton and neutron as being colourneutral. And yet there is a force of attraction between them. It is this force that counters the electrostatic repulsion between the positively charged protons in a nucleus, and is responsible for the nucleus sticking together. To understand how this strong force between nucleons comes about, let us recall how atoms form composite molecules in spite of being themselves electrically neutral. This so-called Van der Waal's force arises through the electrons of each atom rearranging themselves so that they are partly attracted to the nucleus belonging to the other atom. Thus is generated an external remnant force capable of binding the atoms together. In the same way, the quarks within a nucleon can adjust themselves in such a way as to produce an external force capable of attracting the constituents of the neighbollring nucleon—this despite each nucleon having no net colour charge. Thus we see that the strong force operating between nucleons is to be regarded as a 'leakage' of the more fundamental gluon force operating between the constituent quarks.

The strong, or gluon force therefore takes its place as one of the different types of force found in nature. As regards the gravitational, electric, and magnetic forces, these are long-range forces and so give rise to easily observable macroscopic effects—planetary orbits and the emission of radio waves, to mention but two obvious examples. The strong force, on the other hand, is short-ranged, acting over distances of only 10^{-15} m—those characteristic of the size of the nucleus. It was the short-ranged nature of this force that made it that much more difficult to unravel.

I want now to introduce you to a further force: the weak force. It is not weak in the sense of its intrinsic strength being less than that of electric and magnetic forces; it appears weak because it operates over an even shorter distance than the strong force: only 10^{-17}m. Despite this restricted range, however, it has an important role to play. For instance, there is a chain of nuclear reactions whereby hydrogen, H, can be converted into helium, He, with the release of energy. They occur in the Sun and are the source of its energy. The weak interaction is responsible for the first of these reactions:

$$p+p \rightarrow {}^2H+e^++\nu_e$$
$$^2H+p \rightarrow {}^3He+\gamma$$
$$^3He+{}^3He \rightarrow {}^4He+p+p$$

where γ is a high energy photon called a gamma ray, ^2H is a deuteron consisting of a proton and neutron, and ν_e is a neutrino.

The weak force is also responsible for the decay of the free neutron:

$$n \rightarrow p+e^-+\overline{\nu}_e$$

where $\overline{\nu}_e$ is an antineutrino.

Incidentally, you might be wondering what all this talk of a 'force' has to do with particles transforming into each other. I should perhaps explain that *whenever* particles affect each other—in any way whatsoever—physicists talk of this as being due to a 'force' or 'interaction'. This applies not only to changes in motion (the day-to-day context in which we think of forces operating), but also to changes to the identity of particles.

As I mentioned earlier, unlike the hadrons, neither the electron nor the neutrino feel the strong force. This is because they carry no colour charge. The neutrino does not even experience the electric force; it carries no electrical charge. The fact that neutrinos nevertheless do take part in interactions with other particles shows that we must be dealing with another type of interaction—the weak force.

We say the e and the v_e are 'electron-type leptons,' having electron-type lepton number +1. Each of these particles has its antiparticle, e^+ and $\bar{v}_e$ respectively, and these have electron-type lepton number −1. This lepton quantum number is conserved in interactions in much the same way as the baryon number, B, is in the case of hadrons—as you can check with the above reactions. Because they share the same lepton number, there is no difference between the e and v_e, as far as the weak force is concerned.

Why do we speak of *electron-type* leptons? Because there are other types of lepton. There is the muon, μ, and its muon-type neutrino, v_μ; and the tau, τ, with its tau-type neutrino, v_τ. These have their own respective type of lepton number, which also needs to be conserved in reactions. In this way we come to think of the leptons as forming three doublets.

The quarks also come in doublets. Just as earlier we said that the proton and neutron formed an isospin doublet (differently charged states of the same particle, the nucleon), so the u and d quarks (from which the p and n are made) form a doublet. The same goes for the other quarks: s forms a doublet with c, and t teams up with b.

In fact, there is a link between the quark isospin doublets and the lepton 'weak isospin' doublets. They go together in three generations, as in Table 2.

Like the strong interaction, the weak interaction always conserves quantities such as electric charge, baryon number, and lepton number. However, unlike the strong interaction, it does *not* have to conserve quark flavour. So, for example, the decay of the neutron (u, d, d) into a proton (u, u, d) is due to a d quark changing its

Table 2 *The generations of quark and leptom doublets*

Generation	First	Second	Third	Charge
Quarks	u	c	t	$\frac{2}{3}$
	d	s	b	$-\frac{1}{3}$
Leptons	e^-	μ^-	τ^-	-1
	ν_e	ν_μ	ν_τ	0

flavour and becoming the somewhat lighter u quark—the excess energy being emitted. The same applies to hadrons carrying top, bottom, charm, and strangeness. Soon after their creation in a high energy collision, their t, b, c, or s quark transforms into a lighter quark of different flavour. For instance, the decay of the strange particle, Λ°(s, u, d):

$$\Lambda^\circ \rightarrow p + \pi^-$$

involves the s quark changing into a u quark. This is w-hy it is impossible to accumulate supplies of the new particles; no sooner are they created than they rapidly decay back down to the lightest particles. That is why the matter that makes up our world is almost exclusively made of the two lightest quarks, u and d, together with the electron.

To learn more about the weak force, we must retrace our steps a little. When I first spoke of the various forces in nature, I listed the electric and magnetic forces separately. That was indeed how they were originally viewed—different types of force. It took the genius of James Clerk Maxwell, working in the 1860s, to draw together all the known electrical and magnetic phenomena, and recognise that they could all be explained in terms of a single force—the *electromagnetic* force.

This process of unifying forces was not, however, to stop there. Steven Weinberg (1967) and Abdus Salam (1968), building on earlier work by Sheldon Glashow, were

> 物理学家总是试图找到更高层次的"统一"。

16 The Professor's Last Lecture

able to produce an elegant theory which accounted for the electromagnetic and weak forces in terms of tneir being but different manifestations of a single force; the *electroweak* force.

> 他们可是为此理论的提出而获得了诺贝尔奖。当然，一个前提是接下来讲的后来实验的发现。

For this to be possible, the weak force, in common with the other forces we have been considering, had to be mediated by the exchange of some form of particle. The theory predicted that there would be three of them: the W^+, W^-, and Z^0 At the time, no such particles were known.

The theory was triumphantly vindicated in 1983 by their successful discovery. Like the other new particles, they were found to be unstable, decaying for example in the following ways:

$$W^- \to e^- + \overline{\nu}_e \quad \text{or} \quad Z^0 \to \overline{\nu}_e + \nu_e$$

The decays of the Z^0 proved to be particularly interesting. Not only can it decay into ($\overline{\nu}_e + \nu_e$), but also into ($\nu_\mu + \overline{\nu}_\mu$), ($\overline{\nu}_\tau + \nu_\tau$), or any other type of neutrino/anti-neutrino pair that might exist beyond the three currently known types. The more decay channels open to it, the quicker the Z^0 will decay. Thus, the lifetime of the Z^0 provides a sensitive means of estimating how many types of neutrino/antineutrino combinations there must be. The measurea lifetim e indicates there are but three types of neutrino—the three we have already, dis-covered. From this it follows there are only three lepton doublets.

Furthermore, because the lepton doublets are grouped with the quark doublets to form generations, it seems reasonable to infer that there are likely to be only three quark doublets In other woras, the number of quark flavours is limited to six. This is important. A disturbing feature of the quarks had been that each newly discovered type was heavier than its predecessors: u (5 MeV); d (10 MeV); s (180 MeV); c (1.6 GeV); b (4.5 GeV); and t (180 Ge V). Heavy quarks mean heavy hadrons to contain them. And the heavier the hadron, the more difficult it is to produce. This had caused concern that there might be flavours we could never learn about because we physically did not have the resources to produce them. (How big a synchrotron can one build before the entire grass national product of the planet has been swallowed up in the high energy physics budget?!) However, thanks to the $Z°$, this has ceased to be a problem; we now have good grounds for believing there to be only the six already-known flavours.

Thus, the inventory of elementary particles looks like this:

(i) Six quarks and six leptons;

(ii) Twelve intermediary particles, made up of eight gluons,

the photon, $W^±$, and $Z°$.

> 也算是总结吧。

Thus we arrive at the so-called *Standard Model* of particle physics—a theory that summarises all we have been saying about the constituents of nature and the forces between them. It is a crowning achie-vement. All experiments carried out to date are in agreement with it.

What of the future?

> 谁又能有把握地预言呢？但这也不妨试着猜想，猜错猜对至少现在不知道。

An important line of investigation concerns the unification of forces. Just as the electric and magnetic forces were united, and the resulting electromagnetic force was later united with the weak force, perhaps the electroweak force and the strong force will one day come to be recognised as different manifestations of a common interaction. It has been found that as one goes to higher and higher energies, the strengths of the strong and weak force reduce, while that of the electromagnetic force increases; they appear to be converging. According to currently favoured theory, all these forces become of comparable strength at an energy of around 10^{15} GeV. Should this prove to be the case, we would then know that we are dealing with a single *grand unified force*. (I apologise for the over-the-top name, but that is what it is called!)

One problem is that 10^{15} GeV is an energy we

can never hope to produce in the laboratory (the synchrotron would be just too large). At present the limit on energy we can achieve is 10^3 GeV. But all is not lost. Though such high energy conditions are unattainable, there are expected to be residual effects manifest at ordinary energies.

For instance, one suggested theoretical scheme would result in the proton decaying over a long period of time, via the mode

$$p \rightarrow e^+ + \pi^0$$

Searches are being made for such signs of instability in the proton. But none has been found to date. Nevertheless, proton decay is thought to be one of the ways we might be able to explore aspects of grand unification without having to reproduce ultra high energy conditions.

I should, however, point out that, although we are unable to achieve such conditions in the laboratory, there was a time when those conditions were manifest. I refer to the state of the Universe immediately after the instant of the Big Bang. At that time, the Universe consisted of a dense mix of fundamental particles, moving about at random, and colliding with each other. The temperature was enormousmeaning that the particle collisions were characterised by exceptionally high energies—the kind of energies we have been talking about.

Thus, we envisage an early state of the Universe ('early' meaning around 10^{-32} s!), in which the temperature was 10^{27} K and the particle energies were 10^{15} GeV. At that time, the strong, electromagnetic, and weak forces were of equal strength. As the Universe expanded, it cooled. There was now less energy available to the collisions and it became more difficult to create the heavier particles. This in turn meant the various forces began to acquire their distinctive natures. We call this 'spontaneous symmetry beaking'.

Let me give an analogy. When water cools below its

指望用现有的方式，不断提高加速器的能量，终究没有尽头，是否花得起钱是一回事，而人力总有极限，也许就是另一回事了。

问题是我们无法重现大爆炸的瞬间，别的不说，倘真能重现，观察者——我们人类——就已不存在了。

freezing point, it undergoes a phase change—ice crystals form. Whereas in the liquid condition, all directions are equivalent, the crystal has well defined axes. This means that on crystallization, it has to pick out certain directions in space for these axes. But there is nothing particular about these directions; the choice is quite arbitrary. A second crystal forming elsewhere in the water will almost certainly adopt some other orientation. So, although the axes are a very obvious feature of the crystal, their directions carry no ultimate significance. They obscure the fact that at the fundamental level, all directions are equivalent; there is complete rotational symmetry. We say that this original symmetry of the water has been randomly, or 'spontaneously' broken; it is now hidden.

The same holds with the forces. As the mix of interacting particles cools, it too undergoes a kind of 'phase change'. The very different features of the strong, weak, and electroweak forces become evident—these being the differences that make the forces appear so distinctive at the low energies that characterise most of our experience. But as I have said, there is nothing profound in these differences;they must not blind us to the underlying symmetry they have in common—that of the grand unified force.

Unfortunately, I see my time is running out. There is so much more I could say. For instance, I have said nothing about the question of why the elementary particles have the masses they do. Another fascinating topic is that of *magnetic monopoles*. As you know, these cannot be created by breaking magnets in half. But that doesn't prevent them being produced by some alternative means. This possibility was first suggested by Paul Dirac, but is now predicted by the grand unification theory.

As for ways of extending the scope of the Standard Model, a theory known as *supersymmetry* looks promising. It raises the question of whether the distinction between, on the one hand, exchanged intermediary particles (such as gluons, photons, Ws and Zs), and on the other, the particles doing the exchanging (quarks and leptons) is as clear-cut as we have presented them.

Finally I ought to mention *superstrings*. This is based on the idea that the fundamental particles—quarks and leptons—although they appear point-like, are not in fact points, but tiny 'strings'. They are expected to be incred-ibly small—no more than 10^{-34} m in length—but impor-tantly, they are not simple points as we have been assuming.

超弦理论可是这些年来热门的理论尝试，只是对其前途，人们仍争议颇多。

As you will appreciate, with these last few topics we are venturing into the realm of speculation. Whether any of them will, in the fullness of time, gain acceptance and become as established as the Standard Model is today, no-one can say. We shall just have to wait and see.

走着瞧吧！

17 Epilogue

尾 声

> 这位伽莫夫的合作者好不容易回过头来讲点日常语言了。

It had been a hot, sunny day—ideal for sitting out in the garden. But evening was now drawing in. With the light fading, Mr Tompkins put down the book he was reading.

'What are you doing? Can I look?' he asked Maud, who was sitting alongside him, sketching.

'How many times have I told you? I don't like show-ing my work until it's finished,' she replied.

'You'll strain your eyes in this light,' he added.

She looked up. 'If you must know, I'm trying out ideas for the sculpture.'

'Which sculpture?'

'The one for the lab.'

'What lab? What are you on about?'

'The lab we visited...' She paused. 'Oh dear. I forgot to tell you. Sorry. It was when you were with the nurse having your dressing put on. I got chatting with this Mr Richter—head of Public Relations. Just passing the time until you came back. I jokingly told him he needed a sculpture for the forecourt—outside the Visitors' Centre. He said he had often thought that himself. I got to telling him about my work. He seemed particularly interested in the scorching—with the blow torch. He thought that might be useful in getting across a sense of high temperatures, high energies, violent collisions—that sort of the thing The sculpture wo-uld have to symbolise the sort of work that goes on there; it couldn't just be any old sculpture.'

> 又是象征。

17 Epilogue

'So, do you mean you've got a commission?' asked Mr Tompkins excitedly.

'Good Heavens, no,' smiled Maud.' It doesn't work like that. I have to do sketches, come up with ideas, work out an estimate. They might try other people. We'll just have to see. He seemed intrigued that I had an interest in physics. He thought it should help me come up with something relevant. And of course he knows who my father is; *that* might help too!' she laughed.

She laid aside her sketchpad. Together they gazed up at the first evening stars.

> 就目前来看，对于慕德，这应还算是明智的选择吧。

'Do you ever wonder whether you did the right thinggiving up physics?' Mr Tompkins asked.

She thought for a moment. 'A visit like that's bound to make one think. Cutting edge of science, and all that. But, no. Not really. Oh I am sure I would have had a great time working in a place like that—all very fascinating and glamorous. But I don't know. Working in big teams—on experiments that have to be designed and carried out over five, six or seven years ... I don't think I have the patience for that sort of thing.'

'I still can't get over how big it was—the accelerator,' Mr Tompkins mused. 'Funny thought that: the smaller the object you want to look at, the bigger the machine has to be.'

> 这真是两个不错的警句：想考察更小的物体，却需要更大的机器；为了了解物质最小的部分，也许不得不去考察整个宇宙。

'I think it's funny that in order to examine the smallest bits of matter, you might have to look to the entire Universe. And vice versa: the key to understanding the Universe is to look at the properties of its smallest constituents.'

'How do you mean exactly'?

'Well all that business about spontaneous symmetry breaking in the early Universe. That's all to do with the inflationary theory—the reason why the density in the Universe is close to critical. You know; I told you about it. Don't say you've forgotten.'

'No, no. I remember *that*. But I'm not sure I get the

connection...' Mr Tompkins looked blank.

She continued. 'Remember what Dad said about the phase change when the forces took on their distinctive characterisics. It was a bit like ice crystals forming.'

He nodded.

'Well, one of the things that happen when ice freezes is that it expands. It was the same with the Universe; as it cooled, there came the phase change, and the Universe went into a state of superfast expansion—what we call 'inflation'—before it settled down to the type of expansion we see today. The inflationary period lasted only 10^{-32} second, but it was absolutely crucial. It was during that time most of the matter in the Universe came into being ...'

'Sorry,' Mr Tompkins interrupted. 'Most of the matter...? But I thought all the matter was created at the instant of the Big Bang.'

'No. Only a little matter existed from the very beginning. Most of it came into being a short while after that instant.'

'But how?'

'Well, you know how energy is given out when water changes to ice—the latent heat of fusion. That's how it was with the inflationary phase change; it too released energy—energy that went towards the production of matter. And what's more, the mechanism for producing that matter was such that it was geared to pro duce exactly the right amount to give critical density. And you know what the significance of critical density is.'

> 可是临界密度却难以确切得知。所以对暗物质的研究又引人关注。

'It governs the future of the Universe,' replied Mr Tompkins. 'The galaxies will slow down to a halt, but only in the infinite future.'

'That's right. So, the key to understanding both the origins of the matter of the Universe, and the long-term future of the Universe, lies in understanding elementary particle physics—the physics of the small. Not only that, but we know that for the density to be critical most of the matter of the. Universe has to be dark matter.

17 Epilogue

What that consists of we don't know as yet. It might be that neutrinos contribute by having a mass; it could be that it partly consists of some unknown massive weakly interacting particles left over from the Big Bang interactions. We can only hope to answer those questions by looking to high energy physics.'

'I see what you mean.'

'And the cross-fertilisation operates in the other sense too. The only way of examining the behaviour of fundamental particles at grand unification energies, is to find out what they were doing in the early Big Bang—during the one and only time in the history of the Universe where those energies were, or ever will be, attained.'

Mr Tompkins thought for a moment.

'It really is extraordinary the way everything is linked together,' he murmured contentedly. 'All the things I have learned over the lecture series are connected: fundamental particles and cosmology; high energy physics and relativity theory; fundamental particles and quantum theory. What an extraordinary world we live in.'

'And you might have added cosmology and quantum physics to your list,' said Maud. 'Remember, quantum physics has its greatest effects on the smallest scales, and the Universe itself began small. Quantum physics was in charge in the very beginning.

'Take the cosmic microwave background radiation. At first sight it appears to be uniform—the same in all dir-ections But not quite. If it were completely uniform it would mean the matter emitting it wonld have been uniform. That can't be right. Without at least a degree of inhomogeneity in the density of the matter distribution, there would be no centres around which galaxies and clusters of galaxies could later form. In fact, there *are* inhomogeneities. They occur at the level of one part in 10^5. Very small, but vital. It is these that set the pattern for the large scale structure of the Universe—in terms of clusters and superclusters of galaxies, and the galaxies themselves.

慕德怎么突然变得如此专业，又如此哲学化了？这可与她前面的形象不符。也许在结尾的对话中，作者只好以她作代言人了。

可怜的女士!

'Now the crucial question is: What governed the distribution of those original inhomogeneities? Well, because of the exceedingly small size of the Universe to begin with, it's thought they must have originated in quantum fluctuations. That would be really fascinating—if it turned out that the patterns of the tiniest quantum fluctuations were reflected in the large scale structure of the entire Universe...'

Her voice trailed off. The sound of gentle snoring coming from the other sun-lounger told her there was no point in continuing.